THE RISE OF THE GREEK SOCIALIST PARTY

THE RISE OF THE GREEK SOCIALIST PARTY

MICHALIS SPOURDALAKIS

ROUTLEDGE
London and New York

First published in 1988 by
Routledge
11 New Fetter Lane, London EC4P 4EE

Published in the USA by
Routledge
in association with Routledge, Chapman & Hall, Inc.
29 West 35th Street, New York, NY 10001

© 1988 Michalis Spourdalakis

All rights reserved. No part of this book may be reprinted or reproduced or utilized in any form or by any electronic, mechanical, or other means, now known or hereafter invented, including photocopying and recording, or in any information storage or retrieval system, without permission in writing from the publishers.

British Library Cataloguing in Publication Data

Spourdalakis, Michalis
    The rise of the Greek socialist party.
    1. PASOK — History 2. Greece — Politics
and government — 20th century
    I. Title
    324.2495′074   JN5185.P/

ISBN 0-415-00499-3

Library of Congress Cataloging-in-Publication Data

ISBN 0-415-00499-3

Printed and bound in Great Britain by
Biddles Ltd, Guildford and King's Lynn

Contents

Acknowledgements
Glossary of Greek Terms
Introduction

1	**Towards an Understanding of Greek Politics: The Prolegomena to PASOK's Development**	12
	The Birthmarks of Greek Socialism	13
	The 1960s – A Period of Crisis and the Movement's Conception	18
	Changes in Clientelism, the Party System and the State	18
	The Crisis – Mounting Contradictions	20
	The Centre Union and *Anendotos*	22
	Anendotos and its Aftermath	28
	Stalemate	33
	Dictatorship	35
	Conclusion	37
	Notes	39
2	**From Resistance to Political Movement 1967-1974**	50
	The Resistance	50
	PAK	51
	Democratic Defence	56
	P.A.M. and the Split in the Communist Party	58
	The Resistance Heritage	59
	The *Metapolitefse*	60
	The Road to the 3rd of September	61
	The 3rd of September	65
	The Response	70
	Democratic Defence Joins In	72
	The Political Climate	74
	The First Electoral Campaign	81
	The Result	83
	Notes	86
3	**The Critical Year of 1975 – The Consolidation of Papandreou's Control**	92
	The Post Electoral Debates	93
	The First Attempts at Control	99
	The Pre-Congress Announced	104

Contents

	The Pre-Congress	108
	The Split: The Beginning of the End	114
	The Response	120
	Conclusion	123
	Notes	125
4	**From Movement of Protest to Leading Opposition Party 1975-1977**	131
	A Conducive Environment – The Government of New Democracy	131
	Changing Politics – Changing Image	135
	The Organization	139
	Regrouping after the Split	139
	Two More Expulsions	143
	The Trotskyists Forced Out	143
	The 1976 Split	146
	Crystallization of the New Political Orientation	153
	The First Panhellenic Conference	153
	Building the Electoral Machine	159
	A Change in Image	162
	The Electoral Result	166
	Notes	172
5	**The Short March to Power Completed 1977-1981**	180
	PASOK Becomes the Unchallenged 'Vehicle of *Allage*'	181
	Construction of the Moderate Image Continues	182
	Parliamentarianism Consolidated	184
	Technocracy Consolidated	186
	The Final Fine Tuning	189
	The Organization	193
	The Electoral Campaign and the Pasokean Victory	199
	The Programme	201
	Allage	204
	The Left	206
	The Campaign	208
	The Result	209
	Conclusion	212
	Notes	213
6	**The Movement in Power**	224
	PASOK in Government	225
	The Economy: From Keynesian Hopes to the Austerity of Despair	225

Contents

Controversial Labour Policies		229
A Unique Foreign Policy		233
The Socialist Legislative Record on Other Fronts		236
Strategy		237
The Socialist State: Harvesting the Results of PASOK's Party Structure		242
The Organization: Dealing with the Fruits of Hyperactivity		245
The Organization as an Appendage to the Government		250
The First Congress: The Degeneration of the Promising Socialist Movement Completed		254
A New Role for the Rank and File		258
An Assessment		262
The Opposition		265
Notes		269
7 **Conclusion**		276
Appendix I	Declaration of the Principles and Aims of the Panhellenic Socialist Movement	288
Appendix II	The Central Committees of PASOK	297
Appendix III	Election Results	303
Bibliography		306

For Helga

Acknowledgements

Any production of knowledge, no matter how modest, is a collective process, since it is inevitably the outcome of the cumulative experience of an individual's social relationships. Consequently, it is not possible to mention the names of all the people who have contributed directly or indirectly both to my development and to this work. I must, however, begin this statement with an acknowledgement of all the friends and comrades with whom I worked politically in the mid-1970s in Greece, first within PASOK, and later within *Sosialistike Poria*. This project owes a debt to all of them since our discussions and everyday confrontation and struggle with the realities of Greek society contributed significantly to the ideas which are reflected in the following pages. Equally, I owe an immeasurable debt to the people who as teachers, fellow students, friends and comrades in and around Carleton University, Ottawa, Canada, embraced me with friendship, hospitality and affection in the first half of this decade.

A number of people in Greece and Canada helped me at specific stages of the project with the collection or organization of the research material, with their encouragement, and/or with their constructive comments. Needless to say the omissions, the mistakes and especially the point of view which this book expresses, are my sole responsibility. I am indebted to Stelios Babas, Gerasimos Notaras, Andreas Polites, the late Asteris Stangos, Nicos Yiakoumelos, Giorgos Tsafoulias, Damianos Vasileiades, Kiki Stamatoyiannis, Spyros Manoliakis, Yiannes Ioannides, Pavlos Karioteles, Spilios Papaspiliopoulos, Elias Nikolakopoulos, Apostolis Apostolopoulos and Ken Young. In addition, from the highly uncooperative official PASOK agencies, the "Press Bureau" was the welcome exception, and I thank them very much, particularly Gregores Poulos who went far beyond the duties of his position in helping me with the collection of PASOK documents.

The Council for European Studies saw fit to award me a predissertation fellowship in 1983 which allowed me to make one of my research trips to Greece at a critical point in my work.

I am grateful to Jane Jenson and Reginald Whitaker whose

Acknowledgements

encouragement and constructive criticisms proved vital in improving the quality and in the completion of this book. No one could ask for better critical commentators upon his/her work.

I must express particular gratitude to my friend and teacher Leo Panitch, whose vital assistance turned the original nascent idea into a real project. But even more importantly, his own involvement in the political concerns implicit in the work, in combination with his intellectual experience and prowess created the best synthesis of theory and praxis possible within an institution such as a university.

Finally, to the extent to which any indispensible contribution can be put into words, I would like to publicly thank Helga Stefansson, a very important contributor to the completion of this project. Not only did Helga survive my not rarely anti-social and inconsistent moods during the past few years but she also managed to remain consistently encouraging each and every time that the project ran up against objective or motivational difficulties. In addition, Helga's help was critical to this book because she assisted me in overcoming my linguistic difficulties, and taught me the valuable lesson that in addition to research and writing, editing is at the heart of the successful completion of any project of this magnitude. For all of this and much more, this book is for her.

Glossary of Greek Terms

allage — Change, the slogan made famous by PASOK in the period prior to its election.

Anendotos — the 1960 Center Union's campaign to oust the right wing government from power. The term is actually an abbreviation of the term "Anendotas Agonas" (Intransigent Struggle).

A.T.A. — Aftomate Temarithmike Anaprosarmoge (Automatic Index Adjustment).

Demokratike Amena (D.D.) — Democratic Defence, a resistance organization during the junta period, and one of the co-founders of PASOK.

E.A.M. — Ethnikon Apeleftherotikon Metopon (National Liberation Front).

E.D.A. — Eniaia Demokratike Aristera (United Democratic Left).

EDEN — Ellenike Democratike Neolaia (Greek Democratic Youth).

EFFE — Ethnike Enose Elladas (The National Federation of Greek Students).

E.L.A.S. — Ellenikos Laikos Apeleftheroticos Stratos (Greek People's Liberation Army).

E.R.E. — Ethnike Rizospastike Enose (National Radical Union).

KE.ME.DIA. — Kentro Meleton kai Draphotises (Centre of Studies and Enlightenment), the theoretical bureau of PASOK.

K.K.E. — Kommunistiko Komma Elladas (Communist Party of Greece).

K.K.E. (esoterikou) — Kommunistiko Komma Elladas (esoterikou) (Communist Party of Greece, interior). i.e. the Euro-Communist Party in the country.

kommotarches — the local representatives of the politicians of various parties.

kladikes — PASOK's professional and sectoral clubs.

kotsabasides — the pejorative name for the local tax collector during the years of the Ottoman domination of the Balkans.

laokratia — people's republic, government by the people.

Glossary of Greek Terms

metapolitefse — the period following the fall of the junta and the summer of 1974.

micro-mesea stromata — lower-middle strata. A term invented by PASOK's leadership in its effort to include all constituencies in its promises.

Nea Demokratia (N.D.) — New Democracy.

PAK — Panelleneo Apeleftherotiko Kinema (Panhellenic Liberation Movement), the resistance movement founded abroad by Papandreou, and one of the co-founding organizations of PASOK.

PASP — Panellenea Agonistike Parataxe (Panhellenic Militant Students Front), PASOK's student front organization.

PASKE — Panellenea Agonistike Syndikalistike Kinise Ergasomenon (Panhellenic Militant Trade Union Movement) PASOK's trade union front.

palaiokommatikoi — the name given to the politicians or political activists who were associated with the clientelistic politics of the pre-junta period.

Panspoudastike — the student front organization of the C.P. of Greece.

rousfeti — political favour.

Symmachia Aristeron kai Proodeftikon Dynameon — Alliance of Progressive and Democratic Forces.

S.S.E.K. Sosialistike Syndikalistike Ergatike Kinese (Socialist Worker's Movement) Created after the 1985-1986 crisis in PASOK.

zymosis — fermentation, a word often used by PASOK's leadership to denote discussion and debate.

Introduction

The rise of the Panhellenic Socialist Movement (PASOK) has been recognized as an "unusual" or even "unique" phenomenon. More importantly however, it is rich in insights concerning the study of the party system of the European South, the study of social democracy, the question of the role of individuals in history and finally the theory of political parties. However, it has not become the subject of systematic analyses. At best, the rise of the Greek Socialist Party has been the focus of essentially journalistic presentations which with a couple of exceptions,[1] never went beyond the conjunctural analyses of a particular aspect of the party.

This book attempts to do somewhat more, specifically, to explain the failure of PASOK to reach its (radical) potential and to fulfill its promises. The Panhellenic Socialist Movement (PASOK) was established in 1974, in the immediate post-Junta period, primarily by a wide variety of radical activists and anti-dictatorship militants, many of whom had gathered outside the country around the charismatic personality of Andreas Papandreou. Entering Greek politics as it did, in a period in which the first political reactions to the world economic crisis were taking the form of a right-wing backlash, PASOK dynamically demonstrated its origins and made its presence felt in a remarkable and decidedly radical fashion.

The Party's fourfold political slogans graphically encapsulate its radical orientation: "National Independence", "Popular Sovereignty", "Social Liberation", "Democratic Procedure". PASOK's strategic goals were the "cessation of the exploitation of man by man" and ultimate "human liberation". In the Greek context, the realization of such a strategy was seen as passing

Introduction

through the mass democratic mobilization of the Greek people which could guarantee "the creation of a polity free of the control or the influence of the economic oligarchy". Since behind "NATO and the American bases there are the multinational monopolies and their local subsidiaries", the country was to withdraw from NATO and remove the U.S. bases.[2] Thus, from the outset, PASOK was to define its struggle for "a socialist and democratic Greece" as "based upon the principles of national independence (which) is the prerequisite for the realization of popular sovereignty, (which in turn) is the prerequisite for social liberation and (which in turn) is the prerequisite for the realization of political democracy."[3]

The Movement's anti-imperialist foreign and defence policies were also connected to its internal economic policies, while the latter was seen as the basis for the achievement of internal and societal democracy. In addition, PASOK took pains to link its short-term goal with its strategic goal of "a structural transformation of Greek society", and on occasion, the Movement made no mistake of reminding its audience and left-wing critics that "Marxism is (its) method of social analysis".[4] Moreover PASOK constantly criticised social democracy and avoided an official affiliation with the social democratic Socialist International. Finally, but equally importantly, PASOK promised to develop its own organization along the lines of a mass democratic and participatory party structure, away from both the centralism displayed by the traditional left as well as the clientelistic patterns of mobilization exercised by the center and right wing parties.

PASOK's electoral strength progressed remarkably quickly from 13.6% of the popular vote and 12 out of a total of 300 seats in the 1974 election, to 25.3% and 93 seats in 1977, to a victorious 48% and 172 seats in 1981.[5] Although this was accomplished within a political framework which went beyond the premises and the problematic of Western European social democracy, this short march to power was accompanied by a significant modification in the Movement's promises.

PASOK's radical potential has not been realized. While in power Papandreou's party has put forward a number of policies and displayed structural characteristics which are not only a dramatic departure from its original radicalism but which also vividly indicate that PASOK has permanently locked the door to the realization of its radical potential. The development of its

Introduction

organization along centralized, personalized and overall autocratic lines, the promotion of a defence policy which does very little to challenge the traditional pro-western orientation of Greece, its efforts to control and contain the social forces, and finally its austerity measures, which began in 1983 and reached unprecedented levels in the fall of 1985, are only some of the indicators that PASOK has become an empty shell of its old radical self.

Many would attempt to explain PASOK's development in terms of the intractable problems of governing in the present period of capitalist crisis. Others would focus more on the inertia of some of the oldest arrangements of the Greek social formation (e.g. clientelism) and the impact which these had on PASOK's development. Although certain value exists in these arguments, the roots of the problem are deeper than this.

This study will demonstrate that PASOK's short march to power was established by the course of autocracy of power concentrated within the Movement's structure, by moderation of the party's programme and practices (finally exemplified by the predominace of electoralism and parliamentarism), and by consolidation of old clientelistic practices. That is to say, this study is about the process of the short march to power and why in fact PASOK is not radical. Thus, we will examine the actual process of PASOK's remarkable march to power, and the reasons behind its lack of radicalism.

This study is not about the problems which PASOK faces in power, nor about the role which the traditional patterns of the Greek party system played in PASOK's overall development. Furthermore, this study is not one of the Greek political economy, its position in world capitalism or its impact on PASOK's development. Finally, and this must be stressed, this is not a general study about the Greek political machine and PASOK's own conventional development along these lines.

What this is, is a study which describes, assesses and explains the contradictions involved in PASOK's remarkably short march to power. It focuses primarily upon party structure, since it is there that the Movement's contradictions found expression and were fought out. After all, political parties are before anything else organizations, and any attempt to study them cannot but give serious consideration to this question. This is more so in the case of PASOK, since its organizational structure became the focus of the conflict between its internal political

Introduction

tendencies. Since none of them held numerical superiority, the battle over the question of the Movement's structure became the key and the motor in its development and its march to power. That is not to say that the policies and political orientation of the Movement were not important, but in the case of PASOK, the actual internal debate and conflict over policies took place implicitly, while the organizational ones predominated.

The question of the organization is also important in understanding the contradictions in PASOK's short march to power, because since it was there that its three internal tendencies, namely the *palaiokommatikoi*, the technocrats and the radicals were contrasted and reproduced. In other words party structure was the location at which the tendencies' advance or demise was first and foremost crystallized. It was PASOK's organization which displayed a remarkable flexibility and unprecedented endurance, at least by Greek standards, to the diversity of these tendencies.

In this context PASOK proved itself rather innovative as it manifested a remarkable two sided political discourse, developing along the lines of a peculiar "two faced" pluralism which displayed one image of the party publicly and another, very different one, internally. The first tended to portray the Movement in moderate, realistic and pragmatic colours, while the other tended to be more radical and concerned with the organization's (never openly denied) radical strategic goals. This peculiar political discourse helped to maintain the organization's internal unity since its diverse membership was able to identify with at least some part of the Movement's practices. In fact this political pluralism was proven to be one of PASOK's main strengths and one of the reasons behind its rapid electoral success.

The articulation of the Movement's diverse origins through this pattern of the two sided political discourse was, however, not always smooth, nor did it guarantee peace within the organization. Not rarely, internal disputes were experienced which in turn resulted in internal crises. PASOK made a habit of solving these crises in an authoritarian fashion which in turn developed into the organization's internal *modus operandi* and moved the organization further from its original radical promises. This type of practice came to be accepted not only within the organization but also to be translated into government prac-

Introduction

tices.

However, the process of PASOK's moderate development, its attraction to a broad spectrum of political tendencies, and its unity in the midst of such a dramatic transformation, have been due in large part to the presence, the personality, and the idiosyncrasies of one man: Andreas Papandreou. Papandreou encapsulated within his charisma, the political tendency of the 1960s towards the simultaneous breaking away from and continuation of the old clientelistic patterns, the technocratic current which originated in the same period and was nourished during the Junta years, and finally the radicalism of the resistance. The claim that "PASOK equals Papandreou" is far from an exaggeration, since this built-in pluralism allowed him to enjoy tremendous flexibility and considerable autonomy.

It would be a mistake however, to assume that Papandreou's position in the party structure was unlimited and unconditional. Thus, although many of the landmarks of PASOK's development appeared to have, and in fact do have, Papandreou's personal imprint on them, this was not independent of the Movement's internal tendencies, which were the direct outcome of the organization's origins. It is in this fashion that the individual's (even the charismatic one's) role is framed by the historical circumstances and limited by their capacity to respond to it. Papandreou's role has been conditioned by his (remarkable) ability to express at any given moment the mood and the concerns of various tendencies without alienating completely any one of them.

Of course this formulation does not mean or imply that the Movement's shift towards moderation and its rapid electoral success was solely the outcome of its internal balance of power, or of Papandreou's own capricious initiatives. Political parties are organizations which exist and and breathe in the midst of the given social formation. Thus, parties do not develop independently of the changes of the social formation and should be examined through this prism, since the latter outlines, conditions and sets the limits of their growth and development. In the case of PASOK, there is no doubt that a number of external factors contributed to the contradictions of the Movement's short march to power. The external factors which will be given special consideration are: the political heritage of the resistance; the organizational patterns and the overall political discourse of the other parties (including both the traditional clientelistic and the

Introduction

Bonapartist tendencies of the Greek party system); the electoral competitions; and the changes within the Greek political economy.

Ironically, although the resistance was overall at the basis of PASOK's original radicalism, it was the former's lack of mass support, its political vagueness, and voluntarism which contributed to PASOK's autocratic development under Papandreou's unquestionable command, and which in fact by allowing the maximum breadth of interpretation of the Movement's politics, hastened its short march to power. The resistance itself lacked a mass base, and thus PASOK's creation was the project of a small number of activists, and not the outcome of a mass political process. In this sense, PASOK's forsaking of its promises of democracy and mass participation in decision making no longer appears so surprising.

In addition, partially due to the extenuating circumstances of resistance activity, the groups and individuals involved had a voluntaristic conception of politics and a tendency to define political strategy in a vague, unfinished and scattered fashion. These characteristics were to become the foundation of PASOK's own articulation of politics and the basis upon which PASOK was able to appeal to such a broad spectrum of the population. They were furthermore, the main reason for the unity of the Movement's political currents, since voluntaristic interpretations of vague politics in turn gave PASOK maximum political flexibility. The Movement became an organization whose "unity" was based upon its "diversity". The latter in turn assisted the gradual concentration of power in the hands of the leader who could satisfy all the tendencies within the organization.

The lack of long standing practices of internal democracy among the Greek parties, was anything but conducive to the realization of PASOK's original promise to create a mass democratically run organization. This unfortunate absence in the Greek party system assisted, on the one hand, the natural dislike which some of the Movement's tendencies (e.g. *palaiokommatikoí*) had for the plan while on the other proved detrimental to those currents which were more committed to a democratic PASOK. The latter not only had no precedent to refer to, but their efforts were actively boycotted by the agents of the existing non democratic party structures.

In this context we should mention the failure of the other

Introduction

parties on the left (especially the C Ps) to develop a comprehensive and convincing criticism of PASOK's moderate organizational and political development. This lack was particularly important since it occured in the almost complete absence of a left intelligentsia, which could create a radical discourse rich enough to offset or even to question the organization's movement away from its original promises.

Moreover PASOK's failure to materialize these promises, and in fact its active participation in the reproduction of some of the traditional organizational patterns, namely clientelism, and Bonapatism is attributable to the incredible durability, if not inertia, of these patterns. In fact clientelism was proven so flexible and adaptable that although PASOK's organizational development seriously undermined it as a main structure of mobilization in its traditional form, it later resurfaced in the form of PASOK's "green guards". Furthermore, the dominant role of Papandreou, which by definition has undermined the Movement's radical promise, is the product at least partially, of the long standing practices customary to all significant political leaders in the Greek party system.

Other factors contributed first, to the favourable advancement of certain of the Movement's tendencies, and subsequently to the modification of its initial radical promise. The early election call in 1974, as well as the actual economics of the electoral competition should be seen as major contributors to such a development. When the first post *metapolitefse* election was called only a few weeks after PASOK's establishment, the Movement was forced to move towards electoral efficiency, which under the circumstances, objectively favoured the pragmatism and the status of the electorally experienced *palaiokommatikoi*. This amounted to a further move away from the Movement's original commitment to develop along the lines of a mass, participatory and democratically defined organization.

The existence of the traditionally strong old and new petty-bourgeoisie, the close ties of the new urban working class with the peasantry and the institutions of small ownership, the large peasantry and finally the lack of a long working class tradition have given an overwhelmingly petty-bourgeois flavour to the Greek art of conducting politics. Thus the overall short-sightedness, the individualism, and the acquisitive conformist attitudes of the Greek society have been transferred to the way Greeks see politics. Such attitudes created a situation in which the

Introduction

Movement's original promise to march to power as a mass democratic party whose policies were the outcome of participatory collective procedures could almost be overlooked or simply considered a "meaningless detail".

Finally, PASOK's "conservatism" once in power, in addition to being a natural outcome of the way in which the organization had come to power, was also conditioned by the harsh reality of having to "deliver the goods". That is to say, PASOK faced both economic and political constraints. These contraints, as we will see, were bound to favour certain tendencies of the organization (i.e. the technocrats and the *palaiokommatikoi*), and handicap others, thus further contributing to its moderation and "conservatism".

To repeat, this is a study of PASOK's short march to power which was accomplished by the Movement's drastic departure from its original radicalism. To us it is first and foremost an internal history of PASOK's organization, sensitive of course to external factors.

Chapter I begins our study with a brief examination of those historical elements vital to an understanding of PASOK, and the limits to its radicalism. In other words we will attempt to clarify and reflect on those elements by which PASOK was formed, and against which it was struggling. We will thus focus upon those historical aspects of the Greek social formation which are essential to an understanding of PASOK's present day petty-bourgeois, clientelistic, and autocratic nature, and of the country's economy and state apparatus.

PASOK, like any other organization is the outcome of the conflicts and the contradictions of the social formation. In Greek society, these conflicts take place in an overwhelmingly petty-bourgeois socio-political environment, which has been created by the origins and/or the presence of the old and new petty-bourgeoisie, the new urban working class and the disproportionately large peasantry. Consequently, we will make brief reference to the role of the peasantry and the petty-bourgeoisie in modern Greek history. In addition, since no understanding of the structure of PASOK can be separated from an understanding of the development of the system of patronage in Greece, we will briefly examine the origins of clientelism in the modern historical period. Although it has survived many changes of government and undergone many changes in form itself, clientelism still survives within PASOK's organization and is re-

Introduction

flected within PASOK's practices as Government. References will then be made to economic and state development in Greece, since they are essential to any analysis of PASOK as Government and to certain factors which directly limit its actions in that capacity. Finally, reference will be made to the period of the civil war and its aftermath, the period of "guided democracy" or perhaps more aptly, the period of repressive stability.

The period from the early 1960s onward will be dealt with in somewhat more detail, since it is there that the Movement was conceived – in the rise of the masses after the crisis of the mid-1960's, and the circumvention of their entrance into politics by the dictatorship. It is this crisis which was critical to PASOK's own creation and development, and to the expectations which it aroused. PASOK's political discourse takes meaning only when it is examined in the context of the discourse developed in the political radicalism of that time.

In chapter II, we will examine PASOK's roots in the resistance movement against the Dictatorship, since it was there that its ambiguities, yet also its potential and promise lay. We will first examine the political and organizational characteristics of PAK, Papandreou's own resistance organization, since it was this Movement which became the backbone of PASOK. Then we will pursue a similar examination of the group Democratic Defence which eventually became the co-founder of the Movement. It is the organizational orientation of this group which will prove most insightful to PASOK's subsequent development, particularly to an understanding of the organization's crisis in 1975. Brief reference will then be made to the resistance activity of the traditionl left and the traits it bequeathed upon the post-Dictatorship political environment. In this context we will focus on the split of the Communist Party of Greece in 1968, which scarred the development of the Greek left and in turn marked the creation and later development of PASOK. These later developments were in fact to contribute significantly to PASOK's rapid growth.

The *metapolitefse*, the transition to civilian rule, and generally the political environment leading to the first post-Dictatorship election will then be given special consideration, as this was the location of PASOK's birth and first steps to adulthood. The debates and the actual process that preceeded the "Declaration of the 3rd of September", (PASOK's founding document), will

Introduction

become our focus, since therein we can identify the first articulation of the Movement's political tendencies and can locate the organization's internal contradictions which are vital to an understanding of the peculiarities of its road to power. Futhermore this examination of PASOK's first political statement will prove useful in analyzing the organization's moderation, since it can constantly be used as a measure of the Movement's departure from its original radical promise. The last part of Chapter II will be devoted to the 1974 election which became PASOK's first political project. This early election call and the young Movement's subsequent participation, were to change the organization's delicate internal balance, and in turn the course and the nature of its future development.

In chapters III and IV, we will turn to the actual evolution of PASOK and focus primarily on the internal conflicts generated and the implications of the way in which these conflicts were resolved for PASOK's subsequent orientation. These internal conflicts and splits of the organization not only initiated an autocratic *modus operandi* in the Movement's structure but also contributed to the organization's overall moderation. It was through these internal developments that Papandreou's position was consolidated as the single spokesperson and in effect the primary source of power within the organization. Furthermore, it was at this point that the Movement crystallised its more moderate orientation, which became evident in the party's new pragmatic image and its growing parliamentarism and electoralism. Finally this section will conclude with the first solid evidence of PASOK's success when in 1977, the Movement doubled its electoral support and became the leading opposition party.

In chapter V, we will examine PASOK's final march to power once its internal structure had been consolidated. Here we analyse the conditions and the preparations for the organization's final assault at power. We will examine how the Movement managed to become the unchallenged representative of a moderate definition of the popular call of *Ilage* (change) and how it prepared itself to move into power. In this context, the definition of the call for change, the role of the rest of the left, and the implications of the 1981 electoral campaign will be discussed, since they help both to put PASOK's victory into perspective and explain its further departure from its radical promise. The chapter will conclude with a closer look at the electoral result through which the Movement's unprecedented short march to power was completed.

Chapter VI will examine the Movement in power. It will document and analyse the effects of PASOK's coming to power in a rapid and in a moderate fashion. Here we will witness the results of the Movement's failure to democratize, its increasing overall conservatism, its growing attachment to technocracy as demonstrated in its controversial policies, and in its impact upon the Greek state. We will see the the problems which PASOK's government faced, and their effects on its internal organization. These clearly gave an advantage to certain tendencies, which disturbed the balance of power and in turn further promoted the Movement's controversial moderation and conservatism. Furthermore, we will examine the leadership's response to the organization's existing electoral hyperactivity and enthusiasm, as it was exemplified by the reproduction of a new form of clientelism in the guise of the "green guards" and in Socialist patronage appointments. Finally, before we turn to the opposition to the Socialist rule and to the future prospects of the party, we will conclude our presentation of PASOK's odyssey of moderation towards power with an examination of the Party's First Congress. This major event in its organizational history seems to have crystallized the permanent separation of Papandreou's party from its original promised radicalism.

Notes

1. See: Christos Lyrintzis, *Between Socialism and Populism: The Rise of the Panhellenic Socialist Movement* Unpublished Ph.D. Thesis, London School of Economics, 1983; and Nikos Kotzias, *O 'Tritos' Dromos tou PASOK* (The Third Road of PASOK) Athens: Synchrone Epoche, 1985.
2. From the "Declaration of the 3rd of September". See appendix I.
3. Ibid.
4. From A. Papandreou's speech on 9 August 1975, in idem., *Apo to PAK sto PASOK* (From PAK to PASOK) Athens: Ladias, 1976, p. 167.
5. See appendix III.

1

Towards an Understanding of Greek Politics: The Prolegomena to PASOK's Development

There is a tendency in the West to associate Greece with a long radical tradition. This is perhaps tied to notions of the liberation struggle which began in 1821 against the domination of the Ottoman Empire. However, the roots of PASOK as a radical political movement are really quite recent. If anything, it is the constraints upon this radicalism, rather than the radicalism itself which can more easily be traced to certain historical elements of the Greek socio-political formation. The pre-configuration of PASOK's clientelistic, autocratic and petty-bourgeois nature and of the problems associated with the nature of the economy and the state apparatus with which PASOK has to deal as Government, can be traced to the country's modern historical period. We will try to examine some of these elements historically, because in order to understand both present day PASOK, and Greece, it is necessary to place them into a deeper historical context.

In the first part of this chapter we will consequently attempt to provide not a history of Greece, but rather a general reflection of those elements of the country's modern history which are most relevant to PASOK's present form. First, attention will be drawn to the role of the country's large peasantry and petty-bourgeoisie. These strata not only played important roles in the country's modern history, but even today, as PASOK internalizes and reflects the social conflict, they remain critical to its internal balance of power and subsequently to its limitations and dynamic. Second, the complex system of clientelism, with its roots in the early 19th century also marked Greece to the present day. Admittedly, it changed and adapted to the new political and economic conditions, but it finally came to be reflected in the clientelistic/paternalistic practices of the Movement. Third, it was in the modern historical period that the Greek state apparatus reached mammoth proportions. As it developed and enlarged, it was in fact to leave its mark upon the framework of the modern state apparatus with which PASOK would have to deal as Government.

Fourth, Greece's economic development throughout the modern historical period, was characterized by a failure to industrialize, a dependence on trade, (both of which were obviously tied to the external nature of the largest section of the bourgeoisie), and in later years by foreign domination. These characteristics of the economy were to colour PASOK's rhetoric as a new radical movement and later to haunt it when it was faced with the difficult task of actually governing. Finally, brief reference must be made to World War II and the immediately subsequent civil war which were not only to scar the Greek political discourse as late as the mid-1970s, but also were to help set the stage for the "guided democracy" of the fifties and the radicalism of the sixties, the latter of which would contribute directly to the conception of PASOK.

The Birthmarks of Greek Socialism

By the beginning of the 19th century, the decline of the Ottoman Empire had resulted in the deterioration of the living standards of the occupied peasants[1] and a growing insecurity on the part of the merchants who could no longer count on the protection of the Empire. These conditions resulted in a coincidence of interests and consequently in a tactical alliance between the huge mass of peasants, the sailors, and Greek expatriot capital, against the Ottoman domination. There were, however, widely differing expectations concerning its outcome.[2] It was the peculiarities of this alliance which led[3] to the definition of this struggle almost exclusively along classless and nationalist lines.[4] This classless nationalism was thereafter to remain permanently on the Greek political scene.[5]

The origins of the modern Greek petty-bourgeoisie can be found among the employees of the monstrosity called the Greek civil service, the clerical employees of the banking and shipping sectors, the independent commodity/service producers and the domestic tradesmen. In the modern historical period, the petty-bourgeoisie played an important role in Greek politics, not only because of its numerical strength, but also because of the autonomy of the political from the direct control of the Greek bourgeoisie. Essentially, the latter were not "interested in conquering the government – thus they left it in the hands of the petty-bourgeois bureaucrats, enjoying the state resources by buying off ambitious military officers and politicians (in order to introduce)...new legal arrangements for the furthering...(of their) economic privileges."[6] Thus the bourgeoisie never actually occupied state power, but preferred to run the country through the petty-bourgeois bureaucrats.[7]

The "modern" liberal democratic institutions and processes which were introduced relatively early in Greece's political development (universal male suffrage was introduced in 1844),[8] essentially constituted a way of modernizing the existing precapitalist patronage system.[9] In the absence of autonomous political organizations in the countryside, the atomized struggle of the villager for survival had in fact, no other means of articulation than patronage. In order to protect himself, the villager had to search for a patron, someone with influence who could exert some control over the peasant's life and interests.[10] "Nepotism was an obligation, not a moral fault."[11] Consequently, local patrons often manipulated electoral behavior by terror or by buying people off. This practice was not simply a choice of the villager but was, in fact, actively promoted by the practices and attitudes of the merchants and the government officials.[12]

It is important to remember that clientelism cannot be understood in class terms alone.[13] Greece is a prime example of this. Local clientelism changed as a result of both its internal contradictions and class conflict. When the move away from pure vertical to horizontal types of organizing took place, it was expressed through a transformation of the existing patronage structures. Government reforms introduced from 1880-1920 to build infrastructure, improve educational institutions and protect industry, in fact set a trend in this direction, as they broadened the process of political participation and started to transform the nature of clientelism. The "monopolistic" patronage of the local notables was gradually succeeded by a more open state and party oriented clientelism.[14] This helped to steer recipients outside the traditional network of the patronage system, and towards some mass mobilizations which resulted in the narrowing of the gap between political practices and class locations. In fact, the bourgeois/military movement of 1909 which brought E. Venizelos to power can be seen as an expression of that trend.[15]

The Greek state of the 19th century was one of over-extended size and practices.[16] In 1884 for example, Greece had 214 civil servants for every 10,000 people. The comparable figure was for Belgium 200, France 176, Germany 126, America 113, and Britain 73. In effect, one-quarter of the non-agricultural population in Greece made its living directly or indirectly from the state until the latter part of the 19th century.[17] The percentage of GNP devoted to public expenditure was also much higher than in any other European country.[18] One of the principal reasons[19] for this over-extended nature of the Greek state was the uneven, unbalanced nature of Greek capitalism which meant that the "political" had to intervene in order to regulate and ease the contradictions which the underdeveloped market mechanism could not solve.[20] It is also important to note that the Greek state was largely autonomous *vis-a-vis* the social structure.[21] This was in fact necessary in order to ensure the peaceful coexistence of petty commodity production and the monopolistic tendencies of ex-patriot capital.

In addition, due to the external location of much of the bourgeoisie, it became the state's duty to establish the legitimization process and to consolidate the bourgeoisie's hegemony in the already strikingly weak civil society.[22] This was done through such measures as the control of the mass media, education, the co-op movement, and control of attempts to organize autonomous free associations. And furthermore, since the Greek state was far from encompassing the totality of the Greek nation, it was the state's duty to unify the nation[23] in the aftermath of the revolution and to meet the expectations of national integration. Finally, the size of the Greek state can be explained by its own tendency to promote patronage. Due to the continued absence of large farms[24] and the low level of industrialization, the only avenue out of their plight for the huge mass of small farmers (other than emigration) was the state apparatus. In fact, thousands of rural people were hired on the basis of patronage appointments.[25] This helped to create an enormous, conservative, inflexible bureaucracy, and simultaneously, a spineless state apparatus.

This direct involvement of the state with the patronage system, along with the underdevelopment of the civil society, contributed to the state's omnipotent image, and to a certain statolatry, which although in slightly different form, is present until today. This helped to preempt the autonomous development of the organizations of civil society, and in fact, turned every public demand or criticism by the masses into an intense, though indirect, challenge of power.[26]

Many students of Greek politics have commented on the passion which Greeks have for politics. Since all the public activities of the country were connected to or influenced by the state, politics did become a way of life.[27] However, "politics meant the struggle over the distribution of the spoils, not over the issues."[28] The result was an "a-political over-politicization". The patronage system, the lack of horizontal organizations, the emphasis on regional differences, and the high degree of political autonomy of the state, had divorced political discourse from almost all references to its social base. Thus at the level of political discourse, you were first a supporter of a local politician and a party of its choice, then a Peloponnesean and then a peasant.

In the 19th century, the Greek economy continued to maintain traits of a pre-capitalist formation, with the rural underprivileged making up 95% of the population in 1894.[29] Yet at the same time, Greece was very much integrated into the international division of labour,[30] as Greek capital dominated trade in the Balkans and the Eastern Mediterranean.[31] However, by the late 19th century, the international economic crisis had forced a segment of the expatriate capital to search for opportunities within Greece's borders. Changes introduced by the Trikoupis(1880-1895) and the Venizelos (1910-1920, 1928-1932) governments encouraged this capital repatriation,[32] and in addition, helped to bring banking and industrial capital closer together (thus creating a strong financial sector),[33] and assisted the creation of a national market.[34] The "Asia Minor Catastrophe" of 1922 pushed Greek expatriate capital even further to reorient its activities within the country. This in fact signalled the beginnings of Greek industrial development,[35] which would come to be concentrated primarily in the processing of food, tobacco, textiles, and the light consumer goods.[36] Overall however, the economy remained primarily based on agriculture.[37]

Following the second world war,[38] given the changes in the international division of labour, Greek capital[39] reoriented its activities even more towards the "fatherland." It was at this time that it turned its attention to shipping,[40] the development of which was to turn Greek capital into a partner in imperialism and to contribute significantly to the grossly unbalanced development of the Greek economy. The reasons behind this move[41] include: Greeks' historical familiarity with the sector and the availability of a relatively cheap but highly skilled labour force; the close ties which existed between shipping and financial capital which led the latter to become an enthusiastic guarantor of the former's financing; and the tremendous benefits which were bestowed upon shipping capital by the Greek state.[42] Furthermore, the still shaken conservative forces had to tie the country to the US dominated West, and shipping appeared to be an easy link which just so happened to coincide with the US's need to control but not to own a tramp and tanker fleet. Finally, it seems likely that the Right's state saw that the development of shipping and the support of the primary sector would promote the expansion of petty commodity production and the service sector in a fashion that would be devastating for the working class. Manufacturing was in fact the sector that suffered the most from this development, and it came to be characterized by fragmentation and low productivity.[43]

Greece entered the war in October 1940,[44] and by the spring of 1941, the country was occupied and divided by the German, Italian, and Bulgarian armies. Most of the liberal and right-wing politicians fled after the King to Cairo and formed a government in absentia. The Communist Party, the People's Democratic Party, and other small left-to-centre groups founded the National Liberation Front (E.A.M.),[45] which, along with its military wing E.L.A.S., became the vanguard of the resistance. It liberated some regions, established local governments, and set up the "Provisional Committee for National Liberation" which was regarded as the legitimate government of the country. Its post liberation plans were the establishment of a "People's Republic" or *Laokratia*. However, with the support of the British, then of the US, and following five years of civil war, the established politicians of the right and the centre managed to bring the powerful people's movement to its knees.[46] The victors of the civil war then began themselves restructuring their hegemony throughout the 1950s, the years of "guided democracy"[47] by means of political oppression, the military and paramilitary structure and close ties to the US, in order to ensure the establishment of an anti-Communist, pro-American order.[48]

The experience of the development of an autonomous political movement of the 1940s broke down the pre-war political alliance. The pre-war right-centre split was replaced by a nationalist(right) – left(democrat) one. This political division is important to keep in mind, since even until the present day, the political discourse revolves around it. As time passed, and political and social contradictions proliferated, the Nationalist – Left split changed gradually into a Right – Democrat division. This is worth noting because the unification and the resurgence of the centre in the early 60's as well as the left/socialist discourse capitalized on this political cleavage.

The 1960s – A Period of Crisis and the Movement's Conception

The first major challenge to the nature of the Greek social formation had thus come only with the second world war and the civil war. However, the outcome of this challenge was the defeat of the radical movement, and the country was thereafter consolidated under US and domestic right wing hegemony. The roots of PASOK lie in fact in the challenge to this hegemony – in the rise of the masses after the crisis of the mid-1960s and the circumvention of the masses' entrance into politics by the dictatorship. It is here that we can locate the conception of PASOK, within the tendencies for overcoming the old party system which were voiced both in the context of the left opposition to George Papandreou's Centre Union Party and in the context of the dissent within the traditional and independent left. Consequently it is here that we will begin telling our story in detail.

Changes in Clientelism, the Party System and the State

The foreign occupation, the resistance, the civil war and the economic policies pursued by the right brought about great social disruptions and changes in the 1950s. Naturally not only the pre-war party alignment, but also the party organization and political clientelism had to adjust to the new conditions. As we have said, the mass exodus from the rural areas, migration and urbanization as well as the experience of the people's movement had loosened the grip of the old patronage system over the population. However, a new type of clientelism became the primary tool of mobilizing the rural population and it was primarily used by the right-wing.[49] This political clientelism, although very similar to the pre-war type, was now combined with political and ideological terrorism. In spite of the fact that it was limited mainly to the rural areas, it was able to define the electoral outcome, since it was always combined with gerry-mandering practices.[50] Thus, Greek politics in the 1950s, as other areas of social life, was characterized by a peculiar mixture of "modernity" and "tradition" between patronage networks and centrally structured political parties.[51]

This situation resulted in the development of two dominant characteristics of the Greek party system: the lack of ideological debates and the development of strong leadership. Patronage had made the political process "discontinuous and issueless". As Legg put it "deputies, ministers and candidates wander(ed) from one group to the next in search of patronage and favour, or because of personality clashes with faction leaders".[52] Thus, with the exception of the left which was forced to the margins, ideology and talking about issues became the least practiced political game. A simple "black and white" mentality, one to which the civil war had contributed generously, summarized by a "them" and "us" split, was infused into the political process. It is a mentality and a cultural tendency which was to remain predominant until today.

Another characteristic of the "new" party system was the importance of the leadership. With the social order recently challenged, with a patronage system incapable of controlling the entire population and a party organization without collective processes, the right had to re-establish and reproduce its dominance by the use of strong political personalities. These personalities had to have the approval of the pillars of the regime – the Crown and the army – and at the same time had to have some popular appeal. The two personalities who played that role were General Papagos, commander in Chief of the national forces during the civil war, and Konstantine Karamanlis, a self-made man, a lawyer and politician from a rural area who became popular as the minister of public works and gained the trust of the power bloc after the direct involvement of the Crown. The style of the two leaders and their long presence in power (1951-1963) introduced to the political process and to party organizations, a practice which as we will see, is not unfamiliar to Greeks today. In his monumental study of that period Meynaud called it "plebiscite".[53] It can largely be summarized as a contempt for parliament as the law-making body, the complete subordination of all the party's deputies to the leader's will, and the practice of political surprise and secrecy.

Another characteristic of Greek politics, which is not only consistent with the pre-war situation and a practice followed until today, but also demonstrates the peculiarities of the Greek social formation, is the physical absence of the bourgeoisie from state power. Although it had tightened its ties to the state through the financing of dominant right wing parties and through supporting the institution of the crown, Greek capital continued to leave the exercise of power to the petty-bourgeoisie. Out of the 230 cabinet ministers who served between 1946-1965, 52.6% were professionals (32.6% of these were lawyers),[54] 16.9% had a military background, 19.1% were from the state bureaucracy, while only 4.3% were from the business community.[55]

The Crisis – Mounting Contradictions

In spite of the high rate of growth during the period,[56] the development which Greece was experiencing was soon to run up against contradictions. First, although the portion of agricultural production in GNP declined from 36% in 1951 to 27.7% in 1960, the population employed in that sector showed little change, moving from 51.9% in 1951 to 48.9% in 1958.[57] Thus, the farmers who in effect constituted almost half of the economically active population were receiving less than one-third of the national product. Secondly, gross inequalities started to appear in the manufacturing sector with the coexistence of the plethora of small units with much larger ones.[58] The coexistence of big capital operations and petty commodity producers assisted the former in lowering wages, and increased the levels of relative and absolute surplus value since the household heavily subsidized the reproduction of the labour force.[59] This condition was putting tremendous pressure on the working class. At the same time it was putting the aspirations of the army of people employed in the retail/distribution sector for growth on hold since the capacity of the domestic market was shrinking.

Thus, the growth and prosperity of monopoly capital in Greece was realized at the expense not only of the working class but also, if not primarily, at the expense of all the middle strata as well as the small holding farmers. In fact the two latter social categories, who participated in production both as exploiters and exploited started to display signs of discontent. Urbanization, the narrowing gap between urban centers and the countryside, as well as their gradual integration into capitalist relations, allowed less room for their paternalistic integration into the political process. These masses were indicating strong signs of moving away from clientelism as their main mobilization path. The mass rallies in support of the Cypriot liberation struggle in the late 1950s and most importantly the advancement of E.D.A. as the leading opposition political party – it received a quarter of the total vote in the 1958 election – were strong positive indicators of such a movement.

Furthermore, on the other side of the fence, the dominant fraction of capital (shipping), was facing some tough decisions. The Cuban crisis and the emergence of a number of liberation movements were indicating the beginning of the decline of US imperialism – or at least the saturation of its expansion. This, in addition to the creation of the European Economic Community led some Greek shipowners towards thinking that an unconditional link with the US was putting constraints on the expansion of their activities. These factors in addition to the lack of any decent industrial base and therefore market for their services pushed them towards a limited reorientation of both their political and economic practices. At the political level they started to reconsider their exclusively one-sided "friendship" with the US and the West in general, while at the economic level at least some emphasis on the secondary sector no longer appeared a useless exercise.

Thus, by the late 1950s and early 1960s we witnessed a peculiar coincidence of interests on the parts of a segment of capital and the popular classes and strata: the common denominator being the political change capable of "rationalizing" the political and economic processes. This was the reason for the decline of the Right, the reappearance of the Centre as a political force and generally all the events of the 1960s which will be examined below. For the moment, it is sufficient to understand the contradictory nature of this tendency. It is the key to an analysis not only of the events which followed immediately thereafter, but also the politics of Greece until the present day. The crisis which resulted from the developments of the 1950s has not yet been resolved in Greece. It caused the emergence of the Centre, the political turmoil of the mid-1960s the thunderstorm of the dictatorship, the revitalization of the Right, and finally it brought PASOK to power. Its resolution is nonetheless, as remote as ever.

Thus, by the very beginning of the 1960s, two clear conflicting tendencies had developed in the power bloc. One was composed of a small fraction of the shipowners, the few industrialists in Greece and the "traditional" petty-bourgeoisie, while the other was composed of the majority of shipowners and the social groups which based their parasitic existence on the state apparatus. The first advocated the modernization of the country's economy along industrial lines, the renegotiation of the country's relations with imperialism, and finally the restructuring of the state's dated structure in such a way as to enable the incorporation of the growing mass movement. The latter tendency within the power bloc was more short-sighted. It sought the continuation of the existing accumulation and political processes - organization of internal markets, preservation and expansion of the accumulation which had taken place abroad, and assistance to industrial development. It also wanted the political process to remain as it was, i.e. concentrated around the Crown and the army while the masses were to stay outside the "House of Power", even if force had to be applied to keep them there.

The Centre Union and Anendotos

The most significant development of Greek politics in the 1960s, as a result of the crisis, was the revitalization of the Centre. In September 1961, all the liberal political forces under the leadership of George Papandreou and with the active involvement of the Americans[60] managed to unite into a new party: "The Centre Union". The C.U. claimed the liberal/Venizelist tradition of the country and was essentially the political expression of the modernizing tendency of the bourgeoisie. Its policies, the political discourse it adopted and its class participation, particularly in its higher ranks,[61] indicated precisely that.

The C.U. entered the arena of politics at an opportune time and capitalized on the groundwork laid by E.D.A. and the small existing parties of the Centre. Two months after its establishment, the C.U. managed to capture one-third of both the popular vote and the seats in parliament in the general election. In fact, it required a series of scandalous initiatives of the most reactionary sort on the part of the right to hold the C.U.'s growing influence to that level.[62] For this reason the 1961 election went down in Greek history as the "black election". However, it was this election which clearly indicated the termination of the right-wing reign and the incapability of the traditional left (E.D.A.) to take advantage of the crisis. E.D.A. in fact lost 10% of the vote, declining from 24.43% to 14.63%. But most importantly, the election demonstrated the growing tendency of the mass movement to move away from the confines of the patronage system and towards more open and direct political mobilization.

The structure of the C.U. is interesting to consider briefly, since at least in retrospect, it can be seen as a forerunner to PASOK's own structure. In addition to indirectly undermining clientelism and the old party structures, the C.U.'s organizational intentions were quite radical for a non-left party in the context of Greek politics in the 1960s. According to its constitution (which was created by George Papandreou himself) and adopted after the approval of the parliamentary caucus,[63] the C.U. was to be organized both vertically (at the prefectural, regional and local levels) and horizontally (into professional organizations, women's and youth clubs). This was obviously done in response to the growing popular discomfort with regard to clientelism. Furthermore, the constitution designated the "Party's Congress" as the highest collective body of the organization. However, the C.U. did not manage to organize its party's congress, and fell very short of its original plans to develop vertically and horizontally.[64] The only admirable exemption to the latter was its youth organization (E.DE.N.) which developed along along much more radical lines than the main corpus of the C.U.

Furthermore, due both to its diverse origin and this failure to organize a congress, the organization was run basically by its leader George Papandreou. "The old man of democracy," as his faithful supporters called him, not only led the party but in fact dominated it fully. According to the party's constitution, he was only accountable to the congress, and prior to its being called, he was to share the political responsibilities of the party with the parliamentary caucus. However, Papandreou rarely complied with, or used, the collective processes, and in fact adopted a style similar to his major adversary, Karamanlis. That is to say he appeared more as a reproductive factor to the Bonapartist elements of Greek politics than in opposition to them. Thus, the C.U.'s organizational practices represented simultaneously the breaking away from and the reproduction of the old party structures, a tendency which was to resurface ten years later within PASOK.

Soon after the C.U.'s election victory, G. Papandreou announced its *Anendotos Agonas* (Intransigent Struggle) against the Right. The obvious purpose was the popularization of C.U.'s policies to pave the way for the Centre's return to power. However its real motivation was made clear by G. Papandreou when he justified it to the King as an attempt to contain popular unrest.[65] During the *Anendotos* campaign, G. Papandreou covered thousands of miles criss-crossing the country, using his charisma, talking to mass gatherings, visiting not only the cities but also the remote areas. "For a whole year, while sophisticated politicians doubted whether the campaign was a winner, the old man was traversing the countryside in a curiously enthusiastic yet almost mythical communion with the people."[66]

As they were expressed by their slogans during the *Anendotos* campaign, the policies of the C.U. did not differ much from E.D.A.'s programmes of the 1950s. These policies were the crystallization of the common denominator of the social alliance which the C.U. was expressing. They advocated economic growth and a more just distribution of income, "democratization" of the country and "national independence". "Democratization" meant the elimination of all the particular un-democratic measures and institutions which were established during and immediately after the civil war. They never went so far as to challenge, even by implication, the position of the monarchy in the state structure.[67] To the C.U., independence in foreign policy meant the pursuance of a more flexible foreign policy. Any questioning of the country's close ties with the West was never even implied. As G. Papandreou himself said later on during his programmatic declaration as Prime Minister (December 1963), it meant in addition to the maintenance of ties with the West, that friendly economic and cultural relations with Eastern Europe and the USSR should be sought.[68] Although it appeared radical in the context of the period, it was a foreign policy which was in the objective interests, as we saw above, of a portion of Greek capital i.e. shipping.

More specifically, Papandreou's party was appealing to almost all sides of the social arena by putting forward something concrete for each one of them. Thus, it appealed to industrial capital through its policies in favour of rational/productive economic growth while the inflow of foreign capital was not threatened. So far as the interests of the shipowners were concerned, the promise to reorient foreign policy served them well. To the new urban working strata, the new petty-bourgeoisie and the old petty-bourgeoisie, even the first of which was flavoured by distinct petty-bourgeois attitudes, Papandreou promised equitable taxation, salary increases in the service sector, and both public and private reforms to the educational system which would make "education free for all".[69] The latter was particularly appealing not only to this class but to almost every other social group. Greek society was one with widespread petty-bourgeois aspirations and values, and education seemed a likely path of upward social mobility. The emphasis on productive economic policies was indirectly a promise to the working class. However, an explicit promise for higher wages was made as well. Furthermore, commitments to reform labour legislation and for the democratization of trade unions were also made.[70] Finally, to the peasants, in addition to general democratic reforms which would relax the stifling environment in the countryside, financial support was promised.

At this point, we should underline the nature of the C.U. radicalism. The party's call for an "honest state", an "honorable electoral system", and "honest elections",[71] in combination with the explicit universal social appeal of the party, and of Papandreou's refusal to accept E.D.A.'s offer of support for his first minority government,[72] defined the radicalism not only as classless, but also as non-left.

However, it was neither the policies of C.U. nor the party itself which were the really novel element in Greek politics, but rather it was the effects of the *Anendotos* campaign itself. *Anendotos* broke away from the old forms of political organization and caused the terminal illness of the 1950's brand of clientelism. Since the exceptional period of the civil war and the leftist organizing in the countryside, Papandreou was the first to address the rural masses as a socially coherent group and to ask for their support as such.[73] *Anendotos* introduced politics into Greece on the basis of specific issues and not through the atomization of politics and their articulation through personal favours and accommodations. Of course, this does not mean that the C.U. abstained from clientelistic practices, particularly as it had in its ranks a good part of the old liberal guard. But objectively its openess to the masses was undermining clientelism, at least in its old form.

Another effect of the *Anendotos* campaign was the undermining of the oppressive state apparatus. Papandreou's campaign managed to bring out into the open the cumulative frustration of the masses by voicing them. In spite of its initial goals to contain the mass movement, the masses soon gained in confidence and in effect actively questioned the legal and semi-legal oppressive institutions such as the para-military and the national guards.[74] Finally, *Anendotos* and its eventual outcome (electoral victory for the C.U.) democratized the ideological discourse to unprecedented levels. The C.U.'s march to power, its victory, and *Anendotos* paved the way for the development of cultural activities and new intensive ideological debates in such a fashion that by the mid-1960s a completely new climate had been created.[75] In summary, *Anendotos* marked the entrance of the masses into politics in all its glory and with all its contradictions.

While the growing mass mobilization was boosting C.U.'s influence, the increasing discontent had the opposite effects on the left of the political spectrum. E.D.A. not only did not manage to take advantage of the situation but it also lost considerable support, declining from 24.43 % of the popular vote in 1958 to 11.8% in 1964. To attribute E.D.A.'s failure solely to rightwing terrorism or to the anti-communist atmosphere in general is to adopt at least an incomplete understanding of the phenomenon. It is our understanding that certain other factors, such as E.D.A.'s own policies and structure, were at least as important.

In the 1960s, E.D.A. continued its moderate policies of the 1950s, which differed very little from those advocated by G. Papandreou's party. E.D.A. talked about reorientation of the economy towards productive activities and a more equitable distribution of income. It also rallied for democratization of political institutions, and criticized the dominant role of the US in the country's politics. It never implied any challenge to the regime and never questioned the role of the Crown in it. However, it was not simply the striking similarity of its policies with those of the C.U., which led to its marginalization (albeit within a reasonably unfavourable political climate), but the fact that it made no effort to differentiate itself ideologically from the Centre.

Furthermore, although E.D.A. was probably the only party without clientelistic ties between its deputies and their constituents, it remained a highly centralized and not particularly democratic party. Discussions were of course held at the rank and file level but they rarely had any effect on the final policies of the party. Political directives and initiatives were always the exclusive territory of the top leadership. This organizational pattern was not only contradictory to the party's policies on democratizing the political system but it was also less appealing to the masses, whose political energies had been suppressed for so long. The C.U.'s open and much more direct approach to the mass movement, although in essence not any more democratic, was creating the impression of being so, and therefore seemed to be more appealing to the masses.

In addition to all these factors, E.D.A. was the victim of the theoretical bankruptcy of the outlawed Communist Party of Greece (K.K.E.) which was operating within the party and in effect controlling it. The articulation of both the politics of "peaceful co-existence" and the "stage theory" was translated (as in many other countries) into attempts to form some kind of a front with the "patriotic" forces of the Centre. The latter was fruitless since the C.U. itself repeatedly defined its *Anendotos* struggle as "two-fronted": against the Right as well as against the the left. E.D.A.'s attempts to copy the C.U. because of its electoral success and the former's inability to distinguish itself in any significant manner from the C.U. had removed any reason for the electorate to choose it over the C.U. Thus for all these reasons E.D.A. not only missed the opportunity to give a left wing dynamic to the growing mobilization but also, almost voluntarily pushed itself to the margins of the political arena. In 1964, E.D.A. voluntarily and without receiving any promises in return, withdrew its candidates from certain constituencies in order to boost the C.U. victory.[76] In this way, E.D.A., the main body of the left movement in the 1960s became simply an appendage of Papandreou's party.

The events of the 1960s did not leave the dominant right-wing party (E.R.E.) untouched. E.R.E., representing the most conservative element of the power bloc, could not resist the tidal wave of the entrance of the masses into politics. Its record in power and its class participation did not help its adjustment to the new circumstances. In the summer of 1963, Karamanlis resigned because of a disagreement which developed around the issue of the royal visit to England. The incident hinged on the role of the Crown in the political system which Karamanlis wanted to constitutionally modify.[77] The right thus lost its unifying leading personality in parliament and given the C.U.'s successes, began to display symptoms of a fatal disease. It lost two consecutive elections (November 1963 and February 1964) and by the mid-1960s appeared completely incapacitated. Thus the political representation of the Right was objectively left to the extra-parliamentary poles of power: the Crown and the military.

Anendotos *and its Aftermath*

Anendotos brought G. Papandreou's party to power by the end of 1963. In spite of the mass mobilization and the radicalism of the campaign, when in power, C.U.'s policies were not by any means destructive of the status quo. After all, E.R.E. and C.U. were essentially bourgeois parties.[78] Their differences were only in the tactics of achieving power and in the articulation of socio-economic tendencies as they were derived from the contradictions of the late 1950s and early 1960s.

Thus in power, the C.U. not only intensified industrialization but also did so by consistently promoting the inflow of foreign direct investment. As Andreas Papandreou, then a junior cabinet minister in his father's government put it "foreign capital was not only welcome but necessary to the realization of our overall targets."[79] Most of this capital was no longer coming from overseas but its origin was primarily European. By 1966, the inflow of West German and French capital had almost counterbalanced American.[80] Most of this capital went to the manufacturing of export goods[81] which became 60% of the country's overall exports.[82]

Towards an Understanding of Greek Politics

The economic orientation of the country and its closer ties with expanding European powers allowed Papandreou to attempt a redefinition of the country's relations with the US. This was not done through the taking of anti-American stands, but rather by attempting to control the agencies of US power within the country. Thus Papandreou, very much like his predecessor K. Karamanlis, but with greater consistency and determination, tried to strengthen his government's political authority. This meant bringing the military under the political control of parliament. The latter in turn, given the close ties between the army and the Crown, meant in effect, the undermining of the monarchy's power and control of the political process.

On other levels Papandreou did carry out some of his promises. He introduced impressive and extended educational reforms and undertook measures which significantly improved living conditions and increased incomes in the countryside. But the most important "reform" which the C.U. brought about was the liberal democratic climate which Papandreou's "populist" style brought to the country. The latter broke the previously existing undemocratic, often terrorist conditions of social mobilization. This development was extremely important since the contradictory nature of the alliance which the C.U. was representing soon reached its limits and there was a lot to protest over and react to.

Papandreou's democratic reforms met with severe reaction. This came primarily from the Crown, whose authority was indirectly undermined by the reforms, but it was soon transmitted to the upper bourgeois elements of the C.U.'s top leadership. In 1965 Papandreou decided to move himself into the Defence Portfolio in an attempt to penetrate the "parliament-proof" military apparatus, and bring it under government control. With the approval of the US, the King refused to approve the cabinet shuffle and Papandreou warned the Crown that this might lead to his resignation. Within one hour the King had nominated a new Prime Minister, dismissing Papandreou who just a year earlier had led the polls with 53% of the popular vote. This came to be known as the Royal Coup of the mid-1960s[83] and as it was seen in retrospect, the fatal blow to an already sick democracy.

The governments which followed Papandreou's dismissal were conscious attempts to split the C.U. and therefore to undermine its intimidating mass support. The split away from the

C.U. was led by the upper bourgeois elements[84] of the party but also (according to various press reports), millions of drachmas were used by the Crown to buy the integrity and the votes of a number of C.U.M.P.s. The Royal coup however, did not have the intended results. The more the party was purged of its bourgeois elements, the more united the peasant, petty-bourgeois, and working class alliance of the party became. Soon the radical masses took to the streets creating one of the deepest hegemonic crises of modern Greek history.

It must be pointed out that the common denominator of the convergence of all these social groups into a powerful radical mass movement was democratic reform. It was not explicit class interests but a more general pro-democratic statement that lay behind the movement of the mid-1960s. However, the spontaneous mass protest which took place in the streets of the major urban centres quickly bypassed the policies of both the C.U. and E.D.A. and got out of hand. The open challenge to the patriotic feeling of the monarchy by appealing to the famous last article of the constitution, and the dominant anti-monarch slogan "the people do not like you: take your mother and (get) out"[85] strongly suggested that the masses were challenging the foundation of the regime itself. At the same time both the C.U. and E.D.A. were calling for constitutional order! For a whole week hundreds of thousands of unguided masses rallied along these lines. By doing so in spite of their class origins, they had touched and shaken the pillars of the existing socio-political order: the Crown, the role of the army and US imperialism which supported them.

Furthermore, the mass mobilization which *Anendotos* generated had a rather interesting impact on the traditional left. E.D.A.'s policies, practices and structure did not remain uncriticized from within. In addition to a Maoist split as a result of the USSR – China break (1963) which obviously was not unusual thoughout the world, a number of factions, tendencies and para-organizations sprang up within the party including "Friends of New Countries" and "New Left". These developments indicated nothing but the efforts and the agony of some sections of the left to find new political expression outside the parameters of the existing tradition.

In the winter of 1962-63, a number of left activists, mainly members of E.D.A. who disagreed with the party's policies, came forward with the "Athens' Manifesto". The Manifesto focused on the need for coordination of radical activities by activists on the left through the building of some kind of an open and flexible coalition, something that E.D.A. could not do. As Mikis Theodorakis who was involved in it, later put it, "our view was that there was a developing mass movement among the people and particularly among the youth that...E.D.A. due to the narrowness of its politics, could not express".[86] With the interventions of the official party however, the Manifesto was soon after dropped. However, the assasination of Gregory Lambrakis (the famous "Z") shortly thereafter sparked the creation of the "Democratic Movement of Youth, Gregory Lambrakis". The organization, although very much under E.D.A.'s influence, managed to develop not only in numbers but also introduced new political practices which were more open to alliances and more community oriented.

The events of July 1965 were so unusual, so out of the realm of any simple protest, that they have led some students of that period to suggest that the country went though a very short period of revolutionary conditions.[87] Regardless of whether one agrees with both the theoretical assumptions and the analysis of these studies, one thing is certain. The radicalized masses tried to articulate something which was not, and could not be articulated by their official political expressions. The failure of the masses to do so is not only to be found in the contradictions existing at the base of this mobilization, but also in the profound lack of a non-organic intelligentsia and/or a party capable of articulating in a positive manner the dynamic of that protest. The resolution of this political unrest was left to the naivete of the liberals and the stubborn reformism of the official left.

Soon after his dismissal, faced with this tremendous mobilization, G. Papandreou announced the beginning of a second *Anendotos*. However, as Marx put it, history repeats itself first as tragedy and then as farce. This second *Anendotos* developed purely as an attempt to contain the radicalism of the masses. Papandreou, ignoring or rather twisting the popular demands for radical changes, called for a return to the old constitutional order. And sadly enough, E.D.A.'s response was similar to that of Papandreou's party. In fact the leadership of these two parties often appeared together on occasions of at least symbolic significance.

In the context of the future development of **PASOK**, the most significant development which emerged from this radical mobilization was the change which came about within the Centre Union. As we have already noted, the more that the upper bourgeois elements left the party, the stronger the popular alliance became. This was an insight into the actual tendency and dynamic of the mass radicalism. Thus it did not take long for a distinct left wing development to take place within the party.

Andreas Papandreou, G. Papandreou's son, who had even prior to this point been involved in disagreements with his father on the party's political tactics, led the move. His relative youth, his prestigious academic background, and most of all, his charisma, were the bases of the rapidly rising appeal of this young leader. Andreas, as his friends called him, managed to gather round him the most radical elements of the party, namely the young M.P.s and its youth organization E.DE.N. This faction of the party, soon to be called the "centre-left" brought a more uncompromising and nationalist rhetoric into the political discourse. For example, the centre-left under the young Papandreou's rhetoric was more explicit about the need for national control over the military and the intelligence service[88] and less respectful of the Crown's authority. Furthermore, the new aspect of this centre-left tendency was its not infrequent cooperation with parts of E.D.A. on various local or sectoral issues.[89] Although this practice never achieved the open endorsement of Andreas, it was extremely significant symbolically, given the "two-front" nature of *Anendotos*. However, the centre-left did not go so far as to actively challenge the main line of the second *Anendotos*, that of return to constitutional order or to dispute the party's overall strategic choices.

Just at about this time, a number of independent left-centre intellectuals and some members of E.D.A. created the Papanastasiou Society[90] crafted after the Fabian experience. The Society aimed at the development of an analysis of Greek society and support for pursuing the radical tendencies of the Centre Union. The Society made it clear from the outset that it was not interested in running in elections, although many of its members contributed to Andreas Papandreou's efforts to organize the centre-left tendency of the Centre Union. Thus objectively the Society became a think-tank for A. Papandreou's political initiatives.

However, despite political and social similiarities between the society and Andreas' people, there was an important difference. The Society people, in addition to being sensitive to matters concerning democratic processes, also had serious reservations about Papandreou's personality. This was something which made the collaboration of the two groups practically impossible. For this reason, A. Papandreou encouraged the establishment of a new organization called "Democratic Bonds". This organization was much less intellectual and elitist in character than the Society. Although it did not plan to run in the elections, it aimed at becoming the base for the creation of a new party which would embody the ideas of the centre-left. The characteristics which both of these groups had in common were a technocratic understanding and practice of politics and a detestation of the old type of clientelist politics (*palaiokommatikoi*).[91] The former was exemplified by their emphasis on studying and intellectual work, by their legalistic critique of the Greek state and of capitalism in general. But again what differentiated the Society from "Democratic Bonds" was its faithful application of democratic processes within the group and the refusal of its membership to become followers of any single individual. It seems that the latter did not coincide with Andreas Papandreou's domineering personality and created some friction between him and these groups.[92] These shadows in the relations between the membership of these groups and A. Papandreou are particularly interesting since in a "mysterious" way they would resurface within PASOK ten years later, and develop into a major crisis.

The centre-left within C.U. was to gradually grow away from the main-stream of the party. It was however the dictatorship which once and for all brought about their divorce. The young Papandreou's faction would develop into a separate resistance organization, which after the 1974 coup would transform itself into PASOK and carry with it all the contradictions of this period of gestation.

Stalemate

The attempt of the Crown to avoid Papandreou's reforms had completely failed. Not only had the Crown not managed to establish a legitimate parliamentary alternative to the C.U. but it had also discredited itself to the point of no return. The mass movement, however immature, and the real social basis for the needed reforms were the main causes behind this failure.

By the mid-1960s, no single agent of power was able to legitimately re-establish the badly injured hegemony. On the one hand parliament was incapable of "re-establishing the constitutional order" which meant reconciliation with the shaken authority of the Crown. Under the circumstances, no parliamentary party could function unless "deep reforms" were introduced. This was something that even conservative leaders of the right had recognized.[93] However, these reforms would have left political power even wider open to the mass movement. In spite of the fact that the latter never had any aspirations to challenge the underlying socio-economic order, but only its political configuration,[94] this was potentially dangerous. Its dynamism and the nature of the political reforms would not have left any agents of real power at the exclusive disposal of the bourgeoisie and its imperialist partners. This had paralyzed any attempts to impose a solution upon the crisis through parliament. Both G. Papandreou and P. Kanellopoulos (the new leader of E.R.E.) had come to that realization, and under the auspices of the King came to a secret agreement for the formation of a common government after the election scheduled for May 1967.[95] On the other hand the radical mass movement was out in the cold and on its own. It was without leadership and without any positive sense of where it wanted to go. The conjunctural nature of the social alliance which composed it crippled the possibility of imposing its own hegemonic order.

The country was in a complete stalemate. The only structure with the capacity to impose its will was the military which in fact had remained on the margin of this political turmoil. This authoritarian solution appeared the only way out of the crisis. To the Americans it almost guaranteed an unchallenged and continuous presence in the country, given their close ties with the military. To the more conservative elements of the bourgeoisie it meant a return to the good old order and a definite move away from liberal experimentation. It was also something that the more liberal and nationally oriented faction of the bourgeoisie, given its dependence upon these conservative elements and the unpredictability of the mass movement, could tolerate.

The dictatorship in Greece thus came about as a result of the vacuum which arose from the liberal experiment of the mid-

1960s and reactions to it. It was a classic case of what Miliband describes as the "replacement of 'bourgeois democracy' by conservative authoritarianism", when the popular movements "far from constituting a genuine threat to the capitalist order were...deeply confused."[96] When the tanks of the colonels rolled into downtown Athens, bypassing the plans of the generals for a Royal coup,[97] not only was the discomfort of the bourgeoisie brought to an end but so too the exhausted agony of the popular masses.

Dictatorship

The bankruptcy of the pre-dictatorship policies of both the C.U. and E.D.A. became depressingly clear from the easy time which the colonels had in establishing their order. The psychological preparation of the masses could at best be translated into confronting the police in the streets but never went so far as to anticipate a confrontation with the army.[98] The naivete of the two parties and their unquestioningly firm belief in the liberal democratic rules of the political game had once more left the people alone, "unarmed" and powerless. Thus, this lack of power of the movement and the correspondence of the nationalism of *Anendotos* with the nationalist rhetoric of the colonels broke the backbone of the mass movement and consequently any possibility for autonomous horizontal/class based modes of organizing.

The colonels called their regime a "revolution" and with no major problems managed to put freedoms and civil rights in an iron "cast".[99] They made it clear from the beginning that, as one of their theoreticians put it, "the revolution was to intensify economic development...which had been destroyed by previous political anarchy."[100] Indeed from the very beginning their policies were aimed precisely at developing and expanding the country's economy through intensifying industrialization without abandoning, in fact on the contrary, by stimulating the traditional forms of accumulation.

The economics of the Junta were not new to Greece. They were precisely the programme which the governments of C.U.(and not the right) had initiated just a few years earlier.[101] In fact, it has been claimed that the first 5 year plan of the military regime was nothing but a duplicate of Papandreou's plan for the 1966-1970 period.[102] In addition, it is rather arbitrary on the part of some writers to attribute the economic ills of the country (the deficit in the balance of trade, the increasing foreign and internal debt of the country, etc.) to the Junta and to assume a deterioration in real incomes because of the regime.[103] The former were results of the structural contradictions of the Greek economy while the latter was simply not true. Economic misery and authoritarian/exceptional regimes do not necessarily go hand in hand.[104] In fact, in the case of Greece, between 1966 and 1971, annual increases in the salaries of wage labourers were averaging 9.8% while the consumer price index was increasing by only 2.1% per annum on average.[105] It is in such facts that an explanation of the "bloodless" coup and an answer to the failure of the resistance to become popular must be sought.

The final crisis within the dictatorship came about as a result of the contradictions of the economic and political development of the country. It was a combination of the unplanned inflow of foreign capital (62% more in the 1967-71 period than in the 1962-66 period), the international crisis of the early 1970s and its inflationary effects, the increase in friction between the domestic bourgeoisie (usually industrial) and its internationally oriented counterpart, as well as the situation in Cyprus.[106] As Poulantzas pointed out, it is obvious that the form of the regime of the dictatorship cannot guarantee the peaceful and harmless resolution of the contradictions between the factions of the bourgeoisie.[107] The crisis in Cyprus and the possiblity of war contributed to the Junta's desire to find a less dangerous vehicle to pass power safely to the politicians. A general military draft in an authoritarian regime in combination with the increasing discontent of the population, appeared to be a very undesirable position for the regime to put itself into. Thus, it can be said that it was the unpredictability rather than the strength of the "popular factor" which contributed to the final "fall" of the Junta.

At the social level, pressure against the regime was coming from the convergence of interests of various diverse and competing classes and strata. The anti-foreign sentiment could rally the indigenous bourgeoisie, the traditional petty-bourgeoisie, sections of the working class who had a recent peasant background (and were hoping for more rational industrialization and development) and the peasantry which was forced to be proletarianized in the cities or abroad. In spite of this convergence of the popular classes, they did not manage to organize and articulate their voices and have a say in the political change (*metapolitefse*) of the summer of 1974. It was in reality a change imposed from above.

When K. Karamanlis took his place as the new Prime Minister, he knew that these structural contradictions would have to be resolved soon. But he also knew, as did many others, that the solutions to these old contradictions, which had aged since the early 1960s, would have to be fought out under completely new political conditions.

Conclusion

The foregoing analysis is intended primarily to answer those who claim that PASOK is merely the creation of Andreas Papandreou. Papandreou himself has repeatedly claimed that his party is the political expression of the agonies of three generations of struggle: the generation of the radical movement of the '60s; the generation of the resistance; and the generation of the Polytechnic School uprising. He is partially right. The Panhellenic Socialist Movement is that and more.

Even in a country such as Greece in which politics enjoy a reasonable degree of autonomy, political struggle does not occur in a sociological vacuum. PASOK is the outcome of the social contradictions and conflict in Greek society. It is a social conflict in which, given the peculiarities and the dismorphy of Greek capitalist development, a very important factor intervenes: that of the overwhelmingly petty-bourgeois attitudes of the old and new petty-bourgeoisie, the new urban working class, and the peasantry. In other words, PASOK internalizes and reflects a sort of triangular class struggle within which the "petty-bourgeoisie" play a very important role.

Furthermore, it seems to us that an understanding of PASOK's structure cannot be separated from an understanding of the development of the patronage system in Greece. The phenomenon of PASOK's "Green Guards" of the 1980s is not completely distinct from the phenomenon of *Kommatarches*, the local representatives of the parties, a characteristic of Greek political life since the 19th century. The same can be said for the role of Papandreou in the party. The role which the party leaders traditionally held in the Greek party system from Venizelos to Karamanlis, is very important in allowing one to put Papandreou's role in PASOK into perspective. Finally, PASOK's structure cannot be understood in isolation. The historical experiences of both the traditional left (E.D.A. and the C.P.s) before and during the dictatorship, as well as that of the C.U. have played an important role. They have provided the party with models to avoid or copy and with excuses for its not particularly democratic practices.

PASOK's political discourse must also be examined in such a fashion. Its political and ideological discourse takes meaning only if it is examined in the context of the discourse developed within the political radicalism of the 1960s. In addition, the exceptional historical experience which the country underwent during the dictatorship while the rest of the world was experiencing the resurgence of new ideas must be taken into account. The existence of a Greek revolutionary intelligentsia, its idiosyncracies and its weaknesses are something which cannot be left out of the picture in an analysis of this party.

In addition, the examination of PASOK's political discourse, its actual structural articulation and its impact cannot be separated from the characteristics of the political culture in general. This "obsession of the Greek people with politics" or what we have called their "a-political overpoliticization" is vital to an understanding of PASOK's initiatives and more so to an understanding of the reactions which they generate.

Finally, references to economic and state development in Greece are essential pillars in any analysis of PASOK in power. The problems and contradictions which PASOK's government has faced are to a large extent the result of the monstrosity called the Greek state, of the convoluted Greek political economy, of the pressures of the competing factions of Greek capital as well as of the influence which US imperialism historically has exerted on the Greek political process.

All these highlight the context within which our analysis of PASOK's development will take place. Possibly this chapter has gone too far in its attempt to frame the genealogy of this development. However, it is important that not only the writer but also the reader be armed with a clear historical rather than a merely synchronic perspective.

Notes

1. For more on the social organization of the Greek region during the Ottoman occupation see: T. Vournas, *Syntome Istoria tes Ellenikes Epanastases* (Brief History of the Greek Revolution), Athens: Drakopoulos Editions, no date available.

2. For the popular classes, national freedom also meant social, political and individual freedom from a despotic and authoritarian system. Although initially these aspirations were not explicit or universal to the oppressed revolutionaries, they later became so after the liberation in small local uprisings. For example on the island of Andros, in the 1930s, common ownership was identified with justice and a vision of society based on communes was put forward. See for example the declaration written by D. Bali, and quoted as a footnote in G. Karabelias, *Mikromesea Demokratia* (Small – Middle Democracy), Athens: Kommouna Editions, 1982. pp. 15-16.

Obviously, these aspirations were quite different from those of the ex-patriot bourgoisie or the local representatives of the Turks, both of whom contributed large sums of money to the struggle.

3. The civil war during the 1820s between the different revolutionary factions; the introduction of a very liberal constitution (Trizina 1827) and its abolition a year later by Kapodistria, the first Governor; and his assasination in 1831 are cases in point of the uneasiness of this socio-politicial alliance.

4. G. Kontogiorges, (ed.), *Koinonikes kai Politikes Dynames sten Ellada* (Social and Political Forces in Greece), Athens: Exantas, 1977, p. 33.

5. The striking lack of a political organization of any significance among peasants and shepherds at least until the end of the 19th century is a case in point. Their political representation was left to the urban bourgeois and petty bourgeois political parties through the integrative effects of the patronage system. For some of the reasons behind this see G. Dertilis, "The Autonomy of Politics from the Social Contradictions in 19th Century Greece" in G. Kontogiorges, ibid., pp. 56-59.

In addition, the peasants displayed a striking disunity, with local faction that bordered at times on racism. See J.A. Petropoulos, *Politics and Statecraft in the Kingdom of Greece, 1833-1843*, New Jersey: Princeton University Press, 1968, pp. 19-23; V. Filias, *Koinonia kai Exousia Sten Ellada* (Society and Power in Greece), Athens: Synchrona Keimena, 1974, p. 107.

6. A. Rigos, op. cit., p. 192.

7. An important reason behind this articulation of bourgeois interests by the petty-bourgeoisie is that the inadequacies of the civil so-

ciety did not allow for the development of a solid social base for the formation of political organizations. Thus political parties did not manage to stimulate the creation of distinct cultural practices and institutions of their own. This is not quite true for the left wing parties, which however, were not the protagonists in the political arena. For more on the left wing activity see: A. Elephantis, *E Epaggelia tes Adynates Epanastases* (The Promise of the Impossible Revolution), Athens: Themelio, 1979.

8. Greece was the first among European nations to introduce universal male suffrage, though with some modifications, as early as 1844. (Great Britain 1867, France 1848, Belgium 1893, Italy 1913), F. Borella, *Ta Politika Kommata tes Europes ton Deka* (The Political Parties of the Ten European Nations), Athens: Malliares-Paedea Editions, 1983, p. 26.

9. R. Bendix, *Nation-Building and Citizenship*, New York: Anchor Books, 1969, p. 9.

10. M. Serafetinidis, *The Breakdown of Parliamentary Democracy in Greece*, Unpublished Ph.D. Thesis, London School of Economics and Political Science, 1977, p. 144.

11. J.K. Campbell, *Honour, Family and Patronage*, Oxford: Oxford University Press, 1974, p. 257.

12. Ibid.

13. Mouzelis, "Taxike Dome Kai to Systema Politikes Pelatias" (Class Structure and the System of Political Clientelism) in D.G. Kontogiorges (ed.) op. cit., pp. 113-150.

For an example of a class reductionist approach see: Luciano Li Causi "Anthropology and Ideology: The Case of Patronage in Mediterranean Societies", *Radical Science Journal* I, 1975.

14. N. Mouzelis, "Capitalism and the Development of the Greek State" in R. Scase (ed.), *The State in Western Europe*, London: Croom Helm, 1980, p. 260.

15. Venizelos was a politician whose status and impact on the national scene was so important that even today in certain areas, affiliation with the tradition he established generates political power. He represented the unity of the tendencies of capital with the peasants' aspirations for social mobility, whose interests could not have been accommodated by the dated political structures. At the same time, some studies have shown that his opponents belonged mainly to the parasitic and professional classes of the cities. In G. Mavrogordatos, "E Laike Vase ton Kommaton kai e Taxike Syngrouse Sten Ellada tou Mesopolemou" (The Popular Base of Parties and Class Conflict in Greece in the Interwar Period), *Epitheorise Koinonikon Erevnon* (The Greek Review of Social Research) 28, 1976.

16. Konstatine Tsoukalas, *Koinonike Anaptexe Kai Kratos* (Social Development and the State), Athens: Themelio, 1981., see also K.

Vergopoulos. *To Agrotiko Zetema sten Ellada* (The Agrarian Issue in Greece), Athens: Exantas, 1975, p. 41.

17. R.M. Hartwell, "The Growth of Services in the Modern Economy", in C. Cipolla (ed.), *The Industrial Revolution*, London: Fontana, 1973, p. 378, and K. Tsoukalas, "On the Problems of Political Clientelism in Greece in the 19th Century", in D.G. Kontogiorges (ed.), op. cit. pp.89-90.

18. R.M. Hartwell, op. cit. p. 378.

19. On the general theory of the subject see for example: J. Saul, "The State in Post-Colonial Societies", *Socialist Register* 1974; C. Leys, "The 'Over-developed' Post-Colonial State: A Re-evaluation", *Review of African Political Economy*, January – April 1976; W. Zieman and M. Lanzendorge, "The State in Peripheral Societies", *Socialist Register* 1977.

20. On the importance of the petty commodity sector see M. Serafetinidis, op. cit. pp. 87-99.

On the small farm nature of the primary sector see K. Vergopoulos, *To Agrotiko Zetema sten Ellada* (The Agrarian Issue in Greece) op. cit. Part II, ch. A & B, pp. 104-161.

On the state's active involvement in capital accumulation, especially in banking and shipping see M. Serafetinidis, op. cit., pp. 41-45, 72-81; N. Mouzelis, "Capitalism and the Development of the Greek State", in R. Scase (ed.) op. cit., p. 148; K. Tsoukalas, *Koinonike Anaptexe kai Kratos* (Social Development and the State), op. cit.

21. G. Dertilis, "Aftonomia tes Politikes apotes Koinonikes Antitheses sten Ellada tou 169ou Aiona" (The Autonomy of Politics from the Social Contradictions in 19th Century Greece) in D.G. Kontogiorges (ed.), op. cit. p. 50; G. Karabelias, op. cit. p. 119; N. Mouzelis, *Modern Greece: Facets of Underdevelopment* London: Macmillan, 1978, pp. 14-17.

22. N. Mouzelis, "Capitalism and the Development of the Greek State" in R. Scase, op. cit. p. 262.

23. G. Karabelias, op. cit., p. 56.

24. K. Moskov, op. cit., p. 129.

25. For a detailed analysis of the internal structure of the civil service and its significance see K. Tsoukalas *Koinonikoi Anaptexe kai Kratos* (Social Development and the State). op. cit. pp. 98-144.

26. There were often brutal reactions on the part of the state every time popular unrest was developing. Y. Kordatos, *Selides apo ten Istoria tou Agrotikou Kinematos sten Ellada* (Passages from the History of the Peasant Movement in Greece), Athens: Boukoumanis, 1964; and idem., *Istoria tou Ergatikou Kinematos sten Ellada* (History of the Greek Workers Movement), Athens: Boukoumanis, 1972.

27. "Politics was partially at least, an economic activity, a means of securing a livelihood." J.A. Petropoulos, op. cit., p. 60.

28. N. Mouzelis, "Capitalism and the Development of the Greek State" in R. Scase, op. cit. p. 243.

29. M. Serafetinidis, op. cit.

30. Therefore it was not by accident that occasional crises abroad had significant impact on the Greek economy. The main transmitter of these vibrations was Greek ex-patriot capital. If for example we examine the inflow of capital into the country we will see that it corresponds to the various blows that Greek capital suffered abroad: Crimean War 1853-56, Eastern Crisis 1876, the partition of Romelea by Bulgaria 1885, the rising nationalism in Roumania in the late 19th century, Egypt 1952, Istanbul 1955, Congo etc., in K. Vergopoulos, "Eisagoge sten Neo-ellenike Ekonomia" (Introduction to the Study of the Greek Economy), *Monthly Review* (Greek Edition) 36, 37, May 1983, June 1983, pp. 47-50. Also N. Mouzelis, "Capitalism and the Development of the Greek State" in Richard Scase (ed.), op. cit., p. 259.

31. M. Serifetinidis, op. cit., pp. 36-41.

32. These included infrastructure development, improvement of educational institutions, protectionist measures for industry and state coordination of various economic sectors. See Ch. Korizes, *E Politike Zoe sten Ellada: 1821-1910* (Political Life in Greece), Athens: 1974; K. Tsoukalas, *The Greek Tragedy*, Middlesex: Penguin Books Ltd., 1968; N. Mouzelis, "Capitalism and the Development of the Greek State" in R. Scase, op. cit. p. 243.

33. N. Mouzelis, "Capitalism and the Development of the Greek State" in R. Scase (ed.), op. cit. p. 246.

34. Two other contributing factors were: the expansion of Greek territory from 63,211 km^2 in 1912 to 130,199 km^2 in 1922, see K. Tsoukalas, *The Greek Tragedy*, op. cit., p. 20; and demographic changes such as the entrance of over one million refugees following the Asia Minor Catastrophe and the consequent rapid increases in urbanization, see Ch. Korizes, *E Politike Zoe sten Ellada: 1821-1910* (Political Life in Greece: 1821-1910), op. cit. pp. 137-146. Of these settlers, 473,025 were of urban origin and 578,824 came from the countryside. *National Census 1928*, Athens: National Printshop, 1937. Quoted in A. Rigos, "E Politike Ekfrase tes Defteres Ellenikes Democratias" (Political Expression in the Second Greek Republic) in D.G. Kontogiorges, op. cit.

35. Greek industrial production reached unprecedented levels in the 1920s and 1930s. See M. Nikolinakos, *Meletes Pano ston Elleniko Capitalismo* (Studies of Greek Capitalism), Athens: Neo Sinora, 1976, pp. 55 ff; G. Karabelias, op. cit. p. 127; M. Serafetinidis, op. cit., pp. 66-95.

36. M. Serafetinidis, op. cit.; N. Mouzelis, "Capitalism and the Development of the Greek State", op. cit., p. 246.

37. The population involved in agriculture fluctuated between 65%

and 54% (1922-1961), while it has been estimated that 60-70% of the population not occupied in the agricultural sector was working in unproductive activities. K. Tsoukalas, *Exartese kai Anaparogoge* (Dependency and Reproduction), Athens: Themelio, 1977, pp. 197-208.

In 1922, the cultivated land was 1.24-1.40 million hectares. It became 2.7 million in 1938 and reached 3.8 million in 1961. In addition, while in 1928 Greece was producing 700,000 tons of cereals, in 1938 the production figure became 1,700,000 tonsW K. Vergopoulos, "Eisagoge sten Neo-ellinike Economia" (Introduction to the Modern Greek Economy), op. cit., p. 127.

Financial assistance was also made possible through the Agriculture Bank. For more on its role see M. Serafetinidis, op. cit. pp. 132-134, 276-284; K. Vergopoulos, *To Agrotiko Zetema ste Ellada* (The Agrarian Issue in Greece), op. cit., part II, chapter 3.

38. There is very little agreement in the literature concerning Greece's post World War II development. However there are basically two schools of thought. The first, with a main stream orientation, claims (with variations), that Greece is an "underdeveloped" or at best a developing country. The lack of industrial development and the diffused concentration of capital are among the main arguments in support of this thesis. See for example W.O. Candilis, *The Economy of Greece*, New York: Praeger, 1968; X. Zolotas, *Monetary Equilibrium and Economic Development*, Athens: 1964; A. Kanellopoulos, *E Economia Anamesa Sto Kthes kai sto Avrio* (The Economy Between Yesterday and Tomorrow), Athens: Kaktos, 1980.

The second approach, of a more radical and often Marxist orientation, has overemphasized the role of foreign capital, the US influence and again the lack of a strong secondary sector, adopting, with variations, the metropolis-periphery dichotomy. See for example: A. Papandreou, *Democracy at Gunpoint*, London: Penguin, 1973; N. Poulantzas, *The Crisis of Dictatorships: Portugal, Spain, and Greece*, London: New Left Review Publications, 1976; *Monthly Review* 7 December 1972, especially articles by F. Ladis and A. Papandreou" K. Tsoukalas, *The Greek Tragedy*, op. cit.; N. Mouzelis, *Modern Greece: Facets of Underdevelopment*, op. cit.; S. Babanasis and K. Soula, *E Ellada sten Peripheria ton Anaptygmenon Choron* (Greece on the Periphery of Developed Countries), Athens: Themelio, 1978; G. Samara, *Kratos kai Kephalaio sten Ellada* (State and Capital in Greece), Athens: Sinchrone Epoche, 1977; K. Vergopoulos, *Eisagoge sten Neo-ellenike Economia* (Introduction to the Modern Greek Economy), op. cit.; M. Malios, "E Synchrone Phase tou Kapitalismou sten Ellada" (The Present Phase of Capitalist Development in Greece), Athens: Synchrone Epoche, 1975.

It is our thesis that this latter approach has underplayed a number of important characteristics of the Greek political economy and that often their analysis has become the victim of political strategy and practice. To us, theses of this type often create the impression that Greece is a stage for a puppet show which is performed by subversive alien forces. It is an understanding which indicates the naivete of the researcher, turns the study of history into tales of international intrigue and creates an unacceptable alibi for politicians. In fact this approach has led some so far as to equate Greece with Vietnam. See A. Papandreou, "The Structures of Dependency" in *La Monde Diplomatique* (Greek Edition) November 1977; also in P. Papasarantopoulos (ed.), *PASOK kai Exousia* (PASOK and Power), Athens: Paratirites, 1980, p. 196.

39. According to one source, Greek ex-patriot capital was second only to British in investment in Africa during the interwar period. See G. Karabelias, op. cit. p. 128.

40. N. Psyroukis, *Istoria tes Synchrones Elladas* (History of Contemporary Greece), op. cit., especially Volumes II and III; M. Serafetinidis, op. cit.; M. Serafetindis et al., "The Development of Greek Shipping Capital and its Implications for the Political Economy of Greece" in *Cambridge Journal of Economics* Fall 1981, pp. 189-310; V. Choraphas, "To Efoplistiko Kephaleo sten Ellada" (Shipping Capital in Greece) in *Antitheses* No. 15, October 1983; S. Elefthetiou, "Ellenikos Kapitalismos: Provlemata kai Dilemmata" (Greek Capitalism: Problems and Dilemmas) in *Tetradia* 2-3, Fall 1981; J. Petras, "Notes Towards a Definition of Greek Political Economy", mimeo; G. Karabelias, op. cit.

41. M. Serafetinidis et al, op. cit., pp. 292-296; J. Petras, "Notes Toward a Definition of Greek Political Economy", op. cit. pp. 6-8; M. Serafetinidis, op. cit. chapter I, pp. 49-66.

42. In the words of the chairman of the Union of Shipowners, "the shipping industry in Greece was created by Legislative Decree 1687" of 1953. See *Argo* January 1976, p. 66, quoted in M. Serafetinidis et al, op. cit., p. 295.

43. The 15% of the active population employed in this sector in the interwar period grew to only 19.1% by 1961. See ibid., p. 195; J. Petras, "Notes Toward a Definition of Greek Political Economy", op. cit., p. 6.

Contrary to the general principle of the concentration of production, there were 75,000 productive units in the country in the pre-war priod, and 125,000 by 1963. Of these, 104,308 employed 0-4 people. See G. Karambelias, op. cit. p. 33.

44. For the right, the country's defense coincided with the British interests; for the underground C.P., war against fascism at any expense was the third International's policy – N. Zachariades, the imprisoned leader of the C.P. even wrote three days after the Italian attack (31st of October, 1940) "To this war that is directed by the Metaxas government

everyone has to give all his energy without any reservation" quoted in *Voethema yia ten Istoria tou KKE* (Text-Book on the History of the C.P. of Greece), op. cit., pp. 191-192.

45. N. Svoronos, *Episkopese tes Neo-ellinikes Istorias* (Overview of Modern Greek History), Athens: Themelio, 1976, p. 141.

46. It is not the intention of this work to analyse the details of the political events of the 1940s. For critical analyses see: the monunmental work by D. Eudes, *The Kapetanios*, New York: Monthly Review Press, 1972; T. Fitlin, "Counter-Insurgency: Myth and Reality in Greece" in D. Horowitz (ed.), *Containment and Revolution*, Boston: Beacon Press, 1967, pp. 140-181; R.J. Barnet, *Intervention and Revolution*, New York: Mentor, 1972, pp. 119-157; N. Psyroukis, *Istoria tes Synchrones Elladas* (History of Contemporary Greece), Athens: Epikairotita, 1976, Volume I; N. Svoronos, op. cit. For a more main stream analysis see: D.G. Kousoulas, *Revolution and Defeat: The Story of the Greek Communist Party*, London: Oxford University Press, 1965, parts III and IV.

47. These included a variety of legal and quasi-legal state practices. As highlights of these policies we can refer to the introduction and application of the 509/1946 Act under which the Communist Party was banned, appeals to this act had to be heard by regular and mainly by special military courts; the permanent cancellation of certain civil rights of the 1952 constitution, the introduction of the idea of the "Certificates of Social Beliefs" (Act 516/1949). However, the most important expression of these practices was the actual psychological terrorizing of the population especially in the pre-election period. See M. Spourdalakis, *Convergence of Political Parties: The Greek Case*, Unpublished Master's Thesis, University of Manitoba 1980.

There was also considerable manipulation of electoral law during this period and in fact, the electoral system has been an issue of conflict until today. The reason for this is that it has not been constitutionalized and in effect changes according to the will and the needs of the party in power. Especially during the period under consideration, the electoral system was the basis of some scandalous electoral manipulation. For example in the 1952 election the dominant right wing party (Greek Rally) got 49.22% of the popular vote and 247 seats out of 300. In the 1956 election Karamanlis' party (E.R.E.) got 47.38% and 165 out of 300 seats while a coalition of centre parties got 48.15% and only 132 seats. ibid. pp. 65-69. Also J. Meynaud, *Oi Politikes Dynames sten Ellada* (Political Forces in Greece), Athens: Byron, 1974, pp.138-148. Also available in English and French.

48. N. Mouzelis, "Capitalism and the Development of the Greek State" in R. Scase (ed.), op. cit.

49. 74% of the areas, in which representation was based on patron-client relations, were right-wing strongholds. Out of these only one was

an urban centre (Kalamata). K. R. Legg, *Politics in Modern Greece*, Stanford: Stanford University Press, 1968, pp. 159-162, 324-327.

50. Ibid., p. 159; J. Meynaud, *Oi Politikes Dynames sten Ellada* (Political Forces in Greece), op. cit., pp. 50-62.

51. M. Serafetinidis, op. cit., p. 160.

52. K. R. Legg, *Politics in Modern Greece*, op. cit., pp. 141-143.

53. J. Meynaud, *Oi Politikes Dynames sten Ellada* (Political Forces in Greece), op. cit., pp. 235-253.

54. The predominance of lawyers in Greek politics can be explained by the existence of the patronage system and their importance in an economy characterized by a dispropotionately large tertiary sector. Lawyers had/have resources and skills at their disposal such as knowledge of the extremely complicated administrative process which is based on law with inumerable details, exceptions and contradictions; experience in negotiation and compromise; and oratorical skill which renders them capable of functioning within the patronage system in the country-side and the complex bureaucratic structures of the state capital. Furthermore, lawyers' dominance in politics indicates the importance which this stratum had on an economy based on trade, commerce and generally on the unproductive service sector. This excessive involvement of one of the most parasitic vocational groups in Greek politics symbolizes not only the nature of the political economy but most importantly the weight which the petty-bourgeoisie carried/carries as a result of the country's unusual economic development.

55. K. R. Legg, *Politics in Modern Greece*, op. cit., p. 305. For a breakdown of the same statistical data, party by party, see J. Meynaud, *Oi Politikes Dynames sten Ellada* (Political Forces in Greece), op. cit., p. 284.

56. The average annual growth of G.N.P. in constant prices between 1953-1961 was 6.25%W *National Accounts of Greece* 16, Athens, 1967.

57. M. Serafetinidis, op. cit., p. 191.

58. G. Coutsomaris, *E Morphologia tes Ellinikes Viomechanias* (The Morphology of Greek Industry) Athens: Centre of Economic Research, 1963. pp. 59-60.

59. This is one of many cases in which capital takes advantage of existing "traditional" social relations. It would be interesting for example, to examine in this context the role of women in the production process. For obvious reasons the extensive use of unpaid female labour, especially in the agricultural and simple commodity production sectors was vital to capital expansion.

60. It is known that the US opts for existence of more than one political scenario that they have some control over, in those countries in which they are heavily involved. Thus in the June-July negotiations for the formation of Centre Union, a top officer of the US State De-

partment (McGhee) participated actively promoting the idea. J. Meynaud, *Oi Politikes Dynames sten Ellada* (Political Forces in Greece), op. cit. pp. 106-107.

61. The presence of a number of shipowners and industrialists in the top leadership of C.U. is well known (e.g. G. Mavros, K. Mitsotakis, P. Garouphalias, etc.).

62. After the election both C.U. and E.D.A. came out with what they called the "Black Bible" in which the irregularities of the biased election were outlined. In fact, the whole election was organized in a military fashion under the code name "Pericles". See B. Georgoulas. *Oi Nothes Ekloges tou 1961* (The Biased Election of 1961), Athens: 1961.

63. The constitution was approved on September 27, 1962, over a year after the creation of the party. J. Meynaud. op. cit. pp. 277.

64. Ibid. p. 281.

65. M. Serafetinidis, op. cit., p. 292.

66. K. Tsoukalas, *The Greek Tragedy*, op. cit., p. 175.

67. J. Meynaud, *Oi Politikes Dynames sten Ellada* (Political Forces in Greece), op. cit., pp. 292-297.

68. T. Couloumbis, *Greek Reaction to American and N.A.T.O. Influence*, New Haven: Yale University Press, 1966, p. 174.

69. M. Serafetinidis, op. cit., p. 294.

70. Free collective bargaining was in effect almost unknown in Greece. The government was the ultimate arbitrator of wages and salaries, which had to be approved by ministerial decree. This was also the case in situations in which the two parties (employers-employees) had agreed. Furthermore the leadership of the National Federation of Labour was not by any means democratically elected. Through a very complex set of legal, semi-legal and financial patterns, the government managed to control Trade Unionism. C. Jechinis, *Trade Unionism in Greece*, Chicago: Labour Education Division, Roosevelt University, 1967, p. 160. See also T. Katsanevas, *To Synchrono Syndikalistika Kinema sten Ellada* (The Contemporary Trade Union Movement in Greece) Athens: Nea Sinora, 1981. The latter work is also available in English *Trade Unions in Greece: An Analysis of Factors Determining their Growth and Present Structure* Ph.D. Thesis, London School of Economics, 1980.

71. M. Serafetinidis, op. cit. p. 292.

72. G. Papandreou's first victory was during the 3rd of November, 1963 election, but his party had failed to achieve an absolute majority in parliament. Papandreou prefered to go to election in 52 days rather than accept E.D.A.'s offer of support.

73. M. Serafetinidis, op. cit., p. 299.

74. One of the striking incidents of this kind was a five hour battle between 7,000 policemen and demonstrators in Athens (20th of April,1962). Ibid., p. 300.

75. K. Tsoukalas, *The Greek Tragedy*, op. cit., p. 183.
76. J. Meynaud, *Oi Politikes Dynames sten Ellada* (Political Forces in Greece), op. cit., pp. 121-123.
77. Ibid., pp. 115-118.
78. N. Svoronos, *Episkopese tes Neoellenikes Istorias* (Overview of Modern Greek History), op. cit., pp.306.
79. A. Papandreou, *Democracy at Gunpoint*, op. cit., p. 127. The inflow of foreign capital increased very rapidly and to unprecedented levels. In 1960 it was $11.5 million, in 1963 it was $50 million, and in 1966 it was $157.5 million. See N. Mouzelis, "Capitalism and the Development of the Greek State", op. cit. p. 253. These figures become particularly impressive when we consider that US military aid was terminated by 1964.
80. M. Serafetinidis, op. cit., p. 235.
81. M. Serafetinidis et al, op. cit., p. 301.
82. G. Karabelias, op. cit.
83. For an extensive and impressively well researched analysis of the "Royal Coup" see: J. Meynaud, *Politikes Dynames sten Ellada: E Vasilike Ektrope apo ton Koinovouleftismo* (Political Powers in Greece: The Royal Deviation from Parliamentarianism), Athens: Byron, 1974.
84. It is not by accident that people like P. Garoufalias (an industrialist) and K. Mitsotakis (a shipowner) led the split and the "apostates".
85. Article 114 was the last article of the 1952 constitution. It appealed to the patriotic feelings of the people of Greece for the preservation of democracy.
86. M. Theodorakis, interview in K. St.Martin, *Labrakides: Istoria mias Genias* (Lambrakides: The History of a Generation), Athens: Polytypo, 1984, p. 243.
87. N. Psyroukis, *Istoria tes Synchrones Elladas* (History of Contemporary Greece), op. cit. p. 359.
88. A. Papandreou, *Democracy At Gunpoint*, op. cit., p. 176.
89. K. Tsoukalas, "Class Struggle and Dictatorship in Greece" in *New Left Review* 56, July-August, 1969.
90. Named after Alexandros Papanastasiou, a prominent socialist and Prime Minister for a very short period of time during the inter-war period.
91. See the constitution of "Democratic Bonds", in *Anti* 240 3rd of September, 1983, p. 24.
92. R. Eleftheriou, "Ennia Chronia PASOK" (PASOK, Nine Years), *Anti*. 240, 3rd of September, 1983; and interview with G. Notaras, July 7, 1984.
93. See: Karamanlis' interview in *Le Monde*, 29th of November, 1967 quoted as an appendix in M. Genevoix *The Greece of Karamanlis*,

London: Doris Publications 1973.

94. K. Tsoukalas, op. cit.

95. S. Gregoriadis, *E Istoria tes Diktatorias* (The History of the Dictatorship), Athens: Kapopoulos, 1975, volume I, pp. 38-39.

96. R. Miliband, *The State in Capitalist Society*, London, Quarter Books, 1977, p. 247.

97. It has been proven that the Crown was preparing a coup with the co-operation of the Generals (i.e. a Royal coup). The explanation as to why the US secret service endorsed the young officers' action must be sought in both the class origin of the younger officers (they were much closer to the popular classes than the generals who for years had been associated with the corruption of the palace) and the internal structure of the officers' promotions. The stagnation of the younger officers within the lower ranks for years made them perfect candidates for the job. For a more extensive explanation of the latter point see N. Mouzelis, *Modern Greece: Facets of Underdevelopment*, op. cit., chapter 7.

98. J. Katris, *E Gennesis tou Neofascismou sten Ellada* (The Genesis of Neo-fascism in Greece), Athens: Papazizes, 1974, p. 256.

99. There are numerous books dealing with various aspects of the authoritarian character of the dictatorship. For selected references see bibliography.

100. T. Papakonstantinou, *Politike Agoge* (Political Education), Athens: Kabanas Hellas, 1970, pp. 220-221,224.

101. N. Poulantzas, *E Crise ton Dictatorion* (The Crisis of the Dictatorships), op. cit. p. 27.

102. M. Nikolinakos, *Meletes Pano ston Elleniko Capitalismo* (Studies on Greek Capitalism), Athens: Neo Sinora, 1976, pp. 134-137.

103. See for example: J. Pesmatzoglou,"E Economike kai Koinonike Kleronamia tes Diktatorias" (The Economic and Social Inheritance of the Dictatorship), in I. T. Athanasiades (ed.), *24 Jouliou 1974: Epistrophe ste Democratia kai ta Problemata tes* (July 24, 1974: The Return to Democracy and its Problems), Athens: Estia, 1975, pp. 20-36; B. Nefeloudis, *Apomythopoiese me te Glossa ton Arithmon* (Demythologization with the Language of Numbers), Athens: Armos, 1973.

104. N. Poulantzas, *H Crise ton Dictatorion* (The Crisis of the Dictatorships), op. cit., pp. 24-25.

105. From OECD reports. Quoted in N. Poulantsas, Ibid., p. 103.

106. Ibid., p. 141.

107. N. Poulantzas, *E Crise ton Dictatorion* (The Crisis of the Dictatorships), op. cit., p. 141.

2

From Resistance to Political Movement
1967 – 1974

On 3rd of September, 1974, just months after the fall of the dictatorship, Andreas Papandreou announced the establishment of the Panhellenic Socialist Movement which was to become the first mass socialist party in modern Greek history. Although PASOK may have been conceived in the mid-1960s, its immediate roots can be traced to the developments which took place within the dictatorship period. This chapter will present the historical details of the party's gestation in the resistance period and its early development, both of which would imprint the party for years to come. In addition it will detail the political climate, discourse, and the overall response to this radical initiative. Finally, it will conclude with an examination of the first political project of this new movement i.e. the first election in Greece since the traumatic period of the dictatorship, which was itself very important in shaping the Movement's future form.

The Resistance

In spite of the Junta's unexpected takeover, and the consequent lack of preparation among the population, it was not long before a number of resistance groups emerged. However, the resistance never became a mass movement, as it was primarily undertaken by middle class intellectuals and students. Once again, the resistance was to be carried by the left. The right, or at least that part which did not collaborate with the regime, limited its activities to a paper war and to international public relations.

From Resistance to Political Movement

The resistance displayed some characteristics which may prove instructive in an effort to understand the post 1974 political configuration. In addition to their lack of mass support, the resistance groups were characterized by a conscious effort to distance and demarcate themselves from the pre-Junta political parties. This does not mean that their members had no affiliation with the pre-existing parties but from their policies they could by no means be classified within the mainstream of the old parties. In addition, there were vocal as well as tacit efforts for the formation of new political parties distinctively different from, and in opposition to the old ones.

Furthermore, it is worth noting that the resistance (with exceptions of course) was primarily based abroad, mainly in Western Europe. This fact alone would not have been significant if it had not been for the impact of the uprisings of the late 60s on policies of the resistance groups and subsequently on post 1974 politics in Greece. Thus, although the "Greek May" was bound to be circumscribed by inevitable anti-dictatorship overtones, the resurgence of new ideas during the period did penetrate the Greek consciousness, if primarily through the European based resistance acitivists.

In addition to presenting detailed analyses of the Panhellenic Liberation Movement (PAK) and Democratic Defence (D.D.) which were, at least on paper, the co-founders of PASOK, it will also be useful to take a brief look at the other resistance groups. We will emphasize the major groups on the left, since those on the right, (and some very small usually ultra-leftist groups), had virtually no impact on the post-dictatorship party formation.

PAK

Nine months after the coup (16 January 1968), with the direct involvement of the US embassy and the lobbying of some international personalities, the regime freed Andreas Papandreou. A couple of months later, after a short visit to Washington (8 March 1968), where he met with senators Eugene MacCarthy and Robert Kennedy,[1] Papandreou announced the establishment of the Panhellenic Liberation Movement. Many of his old friends, from the old centre-left along with a smaller number of independent left wing students, gathered around this initiative. And from the outset Papandreou's charismatic personality put its seal on the structure and politics of the new organization.

In establishing PAK, Papandreou did not cut ties with the C.U., and remained its representative abroad. His relations with a number of C.U. activists and personalities were not of course the best. But the friction which occasionally developed was only an extension of the tendencies and political differences which had developed within the C.U. before the coup. These difficult relations between the young Papandreou and C.U. continued for at least two years. During the first years of the resistance, the fact that PAK and especially its (in origin) liberal membership believed that international public relations would be enough to destroy the Junta, helped to maintain the link between Andreas' new organization and C.U. Thus, during this period, the C.U. published Papandreou's pre-Junta speeches as anti-dictatorial propaganda[2] and Papandreou himself, although he never attended C.U. meetings, avoided criticism.

As soon as it became clear, however, that the struggle was to be long and that the pressure of foreign governments was not producing the expected results, PAK started to lose the active support of its liberal members, at least in its sections outside the country,[3] and its left wing membership became much larger. With the radical developments of the 1960s in the West (along with the exposure of its leader to the resurgence of radical economics), PAK moved decisively to the left. As Papandreou stated in one of his letters: 1) Any change in Greece cannot be achieved "except through a dynamic armed struggle... and a confrontation with the colonels' regime". 2) The nature of this struggle requires the formation "of a political – military national liberation movement crafted after those of the third world". 3) The aim of this struggle will be "a Democratic Socialist, not a Social-democratic Greece".[4] It was obvious that these theses could never be adopted by the old conservative centre.

PAK's burning of its bridges with the old Centre, resulted not only in a mass withdrawal of the general membership, but also in a simultaneous increase in the "mass" entrance of radical intellectuals and students into the organization. Despite this "purification" process, PAK did not manage to completely overcome its diverse, often contradictory and extremist political discourse,[5] or to develop a common and coherent language. It fluctuated beween ultra leftist statements (e.g. Papandreou calling himself an "anarchist" in the sense that he believed in the eventual disappearance of the state)[6] and the mainstream traditional, practical concerns of the resistance (e.g. PAK of the interior)[7]. But even when the discourse of the organization began to develop into a more coherent whole, its practical expression was not consistent and uniform. The diversity and the contradictory character of the policies of various sections of PAK were not only a result of the diverse social background of its members – something which could have been overcome since we are talking of a group not exceeding a few hundred members – but of the different political-ideological environments in which they operated throughout Europe and North America.[8]

However, in spite of the controversial character of PAK's discourse, some basic characteristics can be deduced from it. PAK was characterized by a radical, unrefined, and often vague eclecticism. Revolutionist romanticism and idealist voluntarism were tendencies which functioned as the common denominator of the organization's policies. These included: the characterization of the resistance as an anti-imperialist liberation struggle; the adoption of a "dynamic" means of opposing the regime; the rejection of social-democracy and the acceptance of Marxism as a tool of analysis and therefore a move away from the anti-communist sentiments of C.U.; the practical distinction between movement and party as an origin of the non-bureaucratic structure of the organization; and finally the adoption of some kind of decentralized socialization (of the means of production) based on self-management as a model for the future society. However, despite the positive aspects of PAK's discourse, for example its attempt to overcome the deficiencies and crisis of the traditional left and its rejection of social-democracy, the organization's conception of socialism remained vague and very legalistic. This can be seen in its major preoccupation with constitutional reform, the restructuring of the state apparatus, along with the vague references to class structure.[9]

In addition, PAK displayed at least one unique characteristic: its leader A. Papandreou somehow remained at the centre of all the debates. His role was articulated in a rather irrational manner within the most sophisticated discussions on self management, internal party democracy or even Leninism.[10] Very much as happened later on in PASOK, Papandreou's charismatic personality appeared to be both the point of reference of the various poles of these debates as well as the connecting link between the diverse tendencies which emerged.

The pivotal, unifying role of A. Papandreou in the organization was, however, not enough to prevent criticisms or bitter friction. PAK was organized in a double sided, parallel network. There was PAK, the underground organization, as well as "Friends of PAK". The latter was a front organization, established a few months after PAK itself, which aimed at legitimizing PAK's policies. Both in Greece and abroad, PAK was run by a national council, whose members' functions and juristiction were obscure or unknown even to "Friends of PAK". In turn, "Friends of PAK" although it had no say in the policies of PAK, developed in many cases into a vibrant organization, with a reasonably democratic structure. It ran a number of seminars and conferences in which rather sophisticated discussions took place and at many of which the favored theme was the articulation of democracy both within the Movement itself and in the projected future society.

Soon an uneasy relationship developed between these two parts of the organization. Friction centered around questions concerning the decision making processes. It was spurred on by a series of arbitrary appointments made by Papandreou and an unknown cast around him, a practice which became an institutionalized habit. "Friends of PAK" had a marked phobia of the *palaiokommatikoi* practices (old fashioned party practices based on clientelism) and PAK's apparent leadership habits did not appear to move decisively away from them.[11] Although criticism of these practices took place, it remained within the framework of the organization and never developed into a split. The extenuating circumstances of the resistance and the search for political efficiency gave PAK's leadership (Papandreou and the National Council he had appointed) the perfect excuse to control and limit such criticisms on internal democracy. In addition, the question of democracy may have seemed less burning due to the regional disparity of the organization; the loose structural ties allowed some room for political flexibility in various sections of the organization. The internal struggle over the organization's structure ended with a seminar in Darmstdat, West Germany in May 1974. There the organization decided on a new democratic constitution, which contained many of the concerns and proposals of the political wing of PAK. However, the implementation of this constitution was delayed by the political change in Greece and the preparation for transcending the organization and becoming a party.

In fact PAK's radical rhetoric not only bypassed the discourse of the traditional left but also had a significant impact on the development of Greek radical politics. However, its actual resistance activity was not particularly impressive.[12] There was a lot of talk and theoretical preparation, and occasional participation in anti-junta activities, but very little concrete action within the country. By the end of 1973, with the expansion of the organization's influence, particularly within West Germany and Italy, there was a lot of pressure for direct action. However, this was to be put off by means of interminable discussions on the relationship between mass movement(s) and political organization(s). As we will see, this is a theme which was to resurface later on in PASOK with controversial consequences.

Another important aspect of PAK's existence was its relationship to other resistance organizations. Although Papandreou's group managed to move away from the old Centre, it at the same time avoided developing smooth relations with the other resistance organizations. It did so either by demanding an unreasonable basis of co-operation (usually the adoption of PAK's full platform), or by indirectly claiming the leadership of the proposed front.[13] In contrast, at the local level, some level of co-operation was occasionally achieved. PAK's systematic undermining of the unity of the resistance can be attributed first to the aspirations of its leadership to appropriate the resistance.[14] Papandreou appeared to be particularly reserved *vis-a-vis* D.D. which had more or less the same origin and base as his organization. This latter point is important because of what happened in 1975 when, as we will see, the core of the old D.D., (at least on paper the co-founder of PASOK) was in effect ousted from PASOK.

In July 1974, shortly after the fall of the Junta, PAK held its last conference in Winterthur, Switzerland. It was decided that the basis of PAK's programme should be preserved, within whatever form the organization would take in Greece. A minority of hard-core functionaries expressed their fear *vis-a-vis* the *palaiokommatikoi* and put forward the proposal that PAK's members should have the upper hand in any future transformation of the organization. The proposal was rejected since the majority felt that the new organization should be based on political principles and not on agreements made behind the scenes.[15]

When Papandreou returned to Greece in the summer of

1974, it was obvious to any penetrating eye that one era was ending and a new one was beginning. It was an era which would lead not only to the creation of the country's first mass socialist party but also to its "short march" to power just seven years later. But no matter how distinctly different these two periods may be, the formation, the evolution, the political discourse, and the practices and the controversies of PAK constitute the birth marks of PASOK. As such they are crucial to any understanding of PASOK's own development. As Papandreou himself put it later on when PASOK was experiencing some organizational turbulence: "PASOK was the necessary development of PAK...(and) if the role and the organizational structure of PAK had been better understood by the members and the activists (of PASOK), very many unfortunate events... could have been avoided."[16]

Democratic Defence

Democratic Defence (D.D.) was established on the morning of the first day of the coup by some prominent members of the "Papanastasiou Society". Recruiting basically from the youth of C.U. (E.DE.N.) but also from left activists who had dropped out of E.D.A., D.D. soon became one of the largest organizations of the resistance. Although its membership never exceeded a few dozen, it was one of the very few resistance organizations which could claim a mass presence. Declaring as its prime goal "the overthrow of the Junta by all possible means".[17] D.D. soon took actions which made the organization known and gained the reputation of heroes for its members. In fact by 1970, almost two thirds of its initial membership had been arrested by the Junta, thus creating the first symbols of the struggle against the regime, and lending moral support to the democratic but largely passive Greek population.

The unity and the effectiveness of D.D.'s resistance activities do not by any means imply an ideological unity within the organization. However, there were two principles which served as D.D.'s ideological linch pins: first, a firm opposition to the corrupt politics of the 1960s, with a conception and practice of politics on a moral and honest basis; and second, a firm belief in democratic procedures both within the organization and in the projected future society – although the latter often took the form of an idealized vision of parliamentary democracy. The organization recognized organized tendencies within it. Although there was a considerable amount of overlap, it is possible in retrospect to distinguish three. These were: an independent Marxist tendency which was mainly based abroad and was influenced by the political concerns of the European new left movement of the late 1960s; a left-center tendency which remained closer to the ideas expressed by the "Papanastasiou Society" in the 1960s; and a third tendency whose pivotal goal and concern was strictly framed within an anti-Junta framework.[18]

In spite of these internal differences D.D. claimed for itself the title of "socialist organization". In its opening declaration, it made clear that the struggle was not only against the Junta but against the forces which had created it by supporting "the internal reactionary forces... i.e. American imperialism".[19] One of D.D.'s main goals, which it actively promoted, was the unity of the resistance against the regime although it articulated its policies in a such a way that it tacitly excluded the right. In fact, it was around the issue of unity that friction developed between D.D. and PAK.

In a fashion similar to that of PAK, D.D. was an expression both of the radicalism which resulted from the Junta experience and of the events which preceeded it. It represented a move away from both social democracy and the heritage of the third international as it was expressed by the C.P.s in the country. The real difference however, between the two resistance organizations, was that D.D. was lacking a leading figure and that its near obsession with democratic procedure made it more removed from traditional parliamentary practices (clientelism). It can be said that D.D. was an expression of the radicalized, middle class white collar professional strata, which despite its contradictions and incoherence, was hoping for the rationalization of Greek society along the vague and often legalistic lines of independent "democratic socialism". However, D.D. had the reputation of being "an army without soldiers" and was left open to accusations of elitism. This latter characteristic, along with the incoherence and the contradictory nature of its members' social background proved to be the group's fatal flaw. Later on, when D.D. entered PASOK (as co-founder) these weaknesses along with D.D.'s political inexperience, became strikingly clear, especially during the events of the 1975 split.

P.A.M. and the Split in the Communist Party

A couple of days after the coup, active members of E.D.A. and the C.P. established the Panhellenic Liberation Front (P.A.M.). From its inception, P.A.M.'s primary goal was the unity of all resistance organizations in order to "oust the dictatorship; re-establish constitutional and democratic freedoms; (secure) the freedom of all parties and organizations;...(and) free elections with proportional representation...to be organized by a government of all parties".[20] To achieve these ends, P.A.M. adopted virtually all expressions of struggle against the regime "from the most decisive to the most reformist".[21]

But the most significant development within the traditional left during this period was the split of the Communist Party. In February 1968, during the twelfth conference of the central committee, the party split over the issue of jurisdiction of its two political bureaus (interior and exterior).[22] The two factions entered a vicious competition for the membership and the approval of the Soviet C.P. Finally, the Bureau of the exterior (the one outside the country) although in a minority, was recognized by Moscow and a rapid political differentiation of the two factions began to take place.

The C.P. of the interior gradually adopted a style, more so in terms of rhetoric than practice, which was more nationalist and often surprisingly moderate. Thus, through P.A.M. they called upon the resistance to orient itself towards the goal of "National Democratic Change",[23] which was not in fact much different from E.D.A.'s programme of the 1950s. Furthermore, they called for "a socialism based on the Greek reality, which capitalize(s) on the positive and negative experience of building socialism elsewhere, avoid(ing) the adoption of foreign models".[24] Although these claims were very appealing since they had been simmering among many left-wing activists well before 1967, they soon became associated with reformist politics due to the way in which the C.P. (interior) was articulating them within the resistance movement. First they called for an effort to politically exploit the contradictions between the King and the Junta, then shortly thereafter they declared that their struggle was exclusively "anti-dictatoric".[25] They also tried to unite the resistance, including the right wingers, which created problems not only for the other left resistance groups, but also for their own party members.

On the other hand, the politbureau of the exterior had trouble reorganizing the party. However with the assistance of the Soviet and other Eastern European parties and by the promotion of a rhetoric more radical than that of the "other" C.P., emphasizing a "democratic and anti-imperialist" struggle, the C.P. soon managed to become the dominant communist organization. It described the country as "industrial – agricultural" (sic) and maintained that the ultimate goal of the left movement was the achievement of a "New Democracy" which would allow economic development on anti-imperialist grounds and pave the road towards socialism.[26] In practice however, the "Party" was not any more radical than its "interior" counterpart. It condemned armed resistance and it was not until 1972 that it made any real contribution to the resistance.

The C.P.'s split had devastating effects on the participation of the traditional left within the resistance. The two parties' preoccupation with their internal affairs made them and many of their members almost completely ineffective. Only "Rigas Fereos", the youth section of the C.P. (interior) and the student front organization of the pro-Moscow party – Anti-E.F.E.E. (Anti – The National Federation of Greek Students), can claim any real contribution (although much later) to the struggle against the Junta.

The Resistance Heritage

Thus, it would be only fair to say that the main weight of the resistance fell on the shoulders of the numerous radical groups from across the left spectrum which mushroomed in and outside the country (e.g. "20th of October", "Aris Velouhiotis", "People's Revolutionary Struggle – L.E.A."). In addition to their dynamic stand against the Junta, their significance stems from the fact that many of the themes and ideas with which they were preoccupied, resurfaced in the post-1974 political discourse. Thus, although most of these groups disappeared after the fall of the Junta, their ideas did not. These included: self-management; solidarity with third world movements; and their concern with the form of decision making within the organized vehicle of change. PASOK's own political language can be seen as the primary example of the articulation of these ideas. This is in fact an important reason why PASOK later became so attractive to a great part of the non-traditional left.

As we already noted, the organized resistance never became a mass movement and as such never really threatened the regime. The 1973 student uprising can be considered the only exception to this rule. In the autumn of 1973 the students took advantage of the regime's attempt to relax some of its repressive measures, and bypassing the initial conservatism of the C.P., they revolted.[27] They occupied the Polytechnic School for three days and were joined by thousands of people. Again the regime had to bring the tanks into downtown Athens in order to crush the revolt. Similar student uprisings took place in other major cities where universities existed and suffered the same fate.

These rebels, who suffered hundreds of dead and wounded, had clearly enunciated the popular tendencies towards the authoritarian experience. They made it clear that "the main prerequisite for the solution of the problems of the people (was) the immediate overthrow of the tyrannical regime of the junta and the simultaneous establishment of popular sovereignty...(which) is intertwined with national independence from foreign interests which for years have been supporting tyranny in our country".[28] Furthermore, the open condemnation of NATO and US imperialism, as well as the appearance of slogans such as "Greece to the Greeks"[29] were clearly voiced and thus anticipated the post-1974 political environment. The Polytechnic School uprising did not achieve much more than the intensification of the friction between the various factions of the Junta. However, it definitely clarified the fact that post-dictatorship politics could not resemble those which had existed prior to 1967.

The Metapolitefse

The failure of the resistance to become genuinely mass and effective put its stamp on the political change of July 1974 (*metapolitefse*). The people of Greece were in effect non participants in the transformation to civilian goverment. When the

Junta transferred its power to the politicians under the extenuating circumstances of that summer (danger of war with Turkey over Cyprus, economic crisis, and general conscription), the popular factor was limited to scattered street celebrations welcoming the goddess of democracy which the new Messiah (K. Karamanlis) had brought with him from Paris.

However, the reasons which forced the Junta to give up its power, created a climate of national emergency and disaster, which, given the absence of any real organized left alternatives, were to be overcome by "national unity" and great compromise. Karamanlis led a goverment of national unity, which was basically composed of the old politicians of E.R.E., some right-wing technocrats, some prominent figures of the old C.U. and a small number of social-democrats in minor portfolios. This government managed not only to secure a smooth "pass-over" but also in effect to paralyse any opposition.[30]

This climate was bound to have an effect on the post-Junta political configuration. When the people are speechless, the demagogues act freely. And when there is no room for a long/natural gestation, early labour and premature birth(s) are bound to occur with all the dangers that these might entail. PASOK's creation was no exception to this rule. PASOK can in fact be seen as the most representative example of this pattern among the parties which appeared on the Greek political scene in the aftermath of the Junta's fall.

Of course, there is no doubt, that PASOK was the creation of the resistance and particularly of PAK. However, these conditions of the *metapolitefse* shaped its subsequent development and characteristics (e.g. the role of Papandreou, the degree of vagueness and flux in its programme and practice etc). For this reason some reference to the political "zymosis", which resulted in the establishment of the "Movement" will be rather useful.

The Road to the 3rd of September

On 16 August 1974 Papandreou arrived in Athens, and made "Kastri" the villa of his father, his residence. In his first statement on Greek soil, Papandreou declared his belief in the "decision of the Greek people to fight for national independence...unrestricted popular sovereignty, social liberation and cultural exaltation" and that his ambition was simply "to contribute to this difficult and uphill struggle".[31] At "Kastri", members of PAK of the interior, with the help of some old C.U. activists, had organized a welcoming ceremony in an explicit attempt to reclaim the leadership of the old center.[32] Papandreou however, without discouraging them explicitly, declared intensive preparation for the formation of a new political organization.

Discussion and preparation started immediately on what eventually came to be known as the "Declaration of the 3rd of September" which became the founding document of PASOK. The whole process took place in three stages. The stages can be defined by the three different groups of activists who participated in them. These eventually developed into three tendencies which can be identified in the organization even today.

In the first stage, a very small number of PAK activists (8-12 people) started to discuss a programme which had been drafted in Germany. In spite of its generalities and vagueness, the proposal contained the main thrust of the radical problematic developed by PAK abroad. The people who participated in these initial discussions can be said to be the core of the potentially genuine socalist wing of the party in formation. They represented the more politically advanced sections of the radicalized white collar workers, whose political vision, however vague, was defined by the rejection of a liberal/capitalist development and frustrated by the backwardness of the dusty old left. While these discussions centered mainly around the tactical question of the presentation of the programme, Papandreou himself adopted an open door policy *vis-a-vis* the activists of various origins who wanted to see him. Numerous personalities of the resistance or simply activists of the *Anendotos* visited Papandreou to "exchange ideas" and be informed on the process of the production of the programme.

A group of these activists, mainly students (who constituted the second stage), soon managed to form a group. Under the name of "students around Andreas Papandreou" they joined the discussion. They had been independent participants in the Polytechnic School uprising and carried a mixed baggage of ideas ranging from left liberal nationalism to left criticism of Leninism. At least implicitly the students made the need to move beyond the traditional left political practices clear. This became evident especially when they focused their problematic around the questions of self-management and the autonomy of the mass movements *vis-a-vis* the party. The entrance of the students into negotiations around the programme of the new party can be claimed to be the first contact of PAK and Papandreou with the radical political tendencies which had developed within the country during the dictatorship. And, since these students started or rather continued to be active in the universities and to recruit among the universities' populations of overly politicized students, they performed the first actual political intervention of the as yet unborn PASOK. The students also felt quite comfortable among the PAK activists. Similarities in political and social background, and age, the tacit criticism of rest of the left, an idealized nationalism and a widespread voluntarism were the bases for this comfortable adjustment.

Problems arose by the end of August when a small group of C.U. politicians entered the negotiations (the third stage – led by Jannis Alevras, later to become leader of the House).[33] The *palaiokommatikoi* put forward objections around some key points of the programme, namely on the issue of the "socializations". Reflecting the limits of the radicalism of the conformist elements of the petty-bourgeoisie and peasantry, they claimed that the programme/declaration should be more specific about both the nationalizations and the "socializations of the means of production." They suggested that concrete industries should be named instead of the general theoretical statement on the meaning and use of socializing production. Although the *palaiokommatikoi* were regarded with great suspicion by the PAK exterior (including Papandreou himself) and the youth, they did receive support from some members of the PAK interior (e.g. Charalambopoulos, Peloponnesios, Touloupas et al.) with whom, in any case, they shared a common political origin.

As result of this disagreement brought forward by the *palaiokommatikoi*, the first serious friction appeared. On the surface, the disagreement did not seem to be of any significance, because PAK and the youth were controlling the process. All the old C.U. politicians could do was to divorce themselves from the process of the creation of the new party. However, a compromise was reached and on the final document a reference was made to the "socialization" of particular sectors. This detail is not as insignificant as it might seem. It is the first indicator of the dynamism of a vibrant movement and a strong indicator of where PASOK was to move subsequently. Furthermore, it can be considered an example *par excellence* of the articulation of class interests at the political level and a lesson in the dilemmas and contradictions which radical/revolutionary politics run into in application.

On the other hand, the radicals (PAK and the youth) had organizational control of an organization which was not yet even born. However, they had not established any contacts with the population at large. And in fact these contacts would be necessary immediately since the spectre of the general elections was hanging over them. This made the voice of the old politicians, although not great in number, very loud and disproportionally influential. Through the mechanisms of clientelism, the old politicians had maintained some contact with the population, which could very well be further developed given their experience at the electoral game. This advantange which the *palaiokommatikoi* held, in combination with the strong voluntaristic element, diffused in the vague and unfinished discourse of the radicals (not uncharacteristic of the social strata they were representing) led to the compromise.

From their point of view, the *palaiokommatikoi* had very good reasons to accept the compromise and stay in the discussions. With their exceptional political instincts, they understood that their political survival could not be achieved simply by revitalizing old political formulas. It had become strikingly clear that the dictatorship had had a significant effect on political attitudes, even among the social classes that the *palaiokommatikoi* tended to represent – the traditional petty bourgeoisie and the peasantry. Last but certainly not least, the old politicians knew that they were lacking leadership – not just any leadership, but a charismatic one, one capable of bridging the gap between the old structures and practices and the new diffused "radicalism". Andreas Papandreou was the ideal candidate for the job. This was something which further boosted his vital role within the organization.

At this point we must note that in addition to some of the non C.P. leftists and "Democratic Defence" Papandreou had also come into contact with a number of Trotskyist activists (Livieratos, Dalavagas, G. Kostopoulos, Lagakos).[34] Papandreou's relation with Trotskyism allegedly goes as far back as the pre-war period, when as a high school student he belonged to an anti-Metaxa Troskyist group. The Trotskyists in fact participated in the debate on the declaration document but did not exert a significant influence on it.[35]

The details of the discussion, arguments, divisions and compromises surrounding the preparation of the "Declaration" of principles of PASOK are particularly instructive in understanding its subsequent development. The friction and compromise during the last days of its gestation set the delicate parameters of the "Movement's" evolution and set the stage for the necessarily ever growing power of its leader. When Papandreou launched the new "Movement" on the 3rd of September, 1974, very few could have predicted any tension arising among the enthusiastic and apparently united activists, or that in only seven years the movement would enter the echelons of Greek state power. However, PASOK's development and success story was to evolve in large part from that initial internal turbulence and those divisions just described.

The 3rd of September

The Declaration of the 3rd of September (see Appendix I), signalled PASOK's entrance into Greek politics in an unprecedentedly radical fashion. Despite the document's vagueness, convoluted language, and ambiguity, it maintained a distinct similarity to PAK's old programme. It definitely bore no resemblance to C.U.'s programmes or even to those of the pre-1967 E.D.A. At certain points the Declaration superseded, at least rhetorically, the radicalism of the traditional left. It will be useful to attempt to illustrate the document's radical nature and consequently that of PASOK, as it made its first official entrance on the Greek political scene.

In its opening statement, the Declaration took pains to identify the "root of the misfortune" of Greeks as the country's dependency. Part of this misfortune was the dictatorship which until that point had been blamed on a handful of army officers. Now according to PASOK, it was one of the results of the country's dependence. "Greece's dependence upon the imperialist establishment of the US and NATO (should stop since) behind NATO, behind the US bases, there are the multinational monopolies and their domestic subsidiaries". It was because of them that PASOK, making the connection or at least allowing a link between dependency and key internal problems, vividly argued that the "...mining of our national interest, the corrosion of popular sovereignty, the economic decline and the exploitation of the Greek wage earners (occurs)". Thus as Papandreou announced the new Movement's 3rd of September Declaration, he affirmed that all this should stop. That was to be achieved through the struggle of the Panhellenic Socialist Movement for a "socialist democratic Greece", a struggle which "...is based upon the principles that our national independence is the prerequisite for the realization of popular sovereignty: popular sovereignty is the prerequisite for social liberation and social liberation is the prerequisite for the realization of political democracy".

The Declaration's radicalism did not end with this general attempt to link foreign policy and defence issues with the Movement's strategic goals (i.e. "National Independence, Popular Sovereignty, Social Liberation, and Democratic Procedure"). It also contained assertions previously unheard of in the Greek context such as "...the exploitation of man by man must be eliminated" and "social and economic equality among the sexes should be fortified". It also went on to describe the required tactical steps for the "social liberation of the Greek working people (which) is identified with the socialist transformation of society." It further called for the "socialization of the financial system in its entirety, (the socialization) of the basic units of production as well as those of the big domestic and foreign traders." In addition, it called for the creation and promotion of co-ops in the primary sector which "will eliminate the middle man who exploits the product of the sweat...of the farmer". It even went on to make specific reference to the necessity "to secure civilized housing for every Greek family...; and to the abolition of private education. PASOK furthermore, did not fail to make a commitment to the "protection of the environment" and to "...the betterment of the quality of life," which in the Greek context of the time were definitely radical.

This brief glimpse at the details of the 3rd of September Declaration should leave no doubt that PASOK entered the arena of Greek politics in a dramatically radical and dynamic fashion. It makes the "Down with Capital" slogan that some of PASOK's most radical recruits chose to promote during the Movement's rallies and which engendered sarcastic responses from the others on the left (who considered themselves far more radical than this upstart movement), seem not as out of place as one might have thought.

The significance of the "Declaration of the 3rd of September" lies primarily in the fact that it became the linch pin for the diversity of social and political groups which found expression in PASOK. Thus, the document, which even today is the point of reference of the Movement, cannot be dismissed, as has been done by the majority of the organized and unorganized Greek left, as a text which says "everything" and therefore often "nothing". On the contrary, its historical significance and organizational strength, we would argue, derives precisely from the fact that it contains "something" for everyone but not "everything" for everyone. In retrospect, we can argue, at no great risk, that the "3rd of September" declaration has framed the struggle which took place among various tendencies within and over the party. All the tendencies: the socialist (mainly PAK and the other left-wingers of the resistance); the technocratic (represented by the academic and intellectual personalities of the left centre as well by the students – although they wanted to be associated with the left): and the conformist (mainly represented by the *palaiokommatikoi*) despite their apparent differences were able to find shelter in the magic discourse which the document initiated.

Thus, the *socialist tendency* was able to find some of its key preoccupations in the declaration. First there was an understanding of the country's situation in terms of dependency and therefore a major concern for national independence. In spite of its nationalist overtones, this concern for independence allowed some room for an analysis along class lines, since the assumed prerequisite for the achievement of "national liberation" was social liberation. Of course, any explicit reference to class was diluted either through indistinguishable references to classes, social strata and the occupational or temporal origins of other groups (e.g. workers, farmers, "small professionals", youth, artisans, etc.), or by reducing bourgeois rule to the abstraction of "capital". This abstraction was completely inappropriate and incompatible with the otherwise popular nature of the document. It is interesting to note that there is no reference to the bourgeois class as such. However we can say that even this circumscribed class analysis, due to its overall vagueness, did not exclude the possibility of further developing the declaration along class lines. This latter point is important to keep in mind if we want to understand PASOK's "seduction" of the left. Furthermore, references to "new" contradictions, the importance of which the "new left" movement had recognized (e.g. women's liberation, pollution) became another attraction for both the ex-activists of the traditional left, which had been ignoring them, and the leftists who had been politicized by the development of these new movements abroad.

The *technocratic tendency* within the party was in fact the most diffused of the three tendencies. It did not really become significant until after PASOK's success in the 1977 election. It too was able to identify a number of its own themes within the document. The document emphasized administrative decentralization associated with the goal of rational administration, restructuring of the economy, and direct state intervention at all levels. Ironically, within the context of Greek politics at the time these could be largely be interpreted as a further conglomeration of the state and as further bureaucratization. Furthermore, the voluntaristic thrust of the document coupled with legalistic notions of change (e.g. legal reforms to reconstitute the military, separate church and state, and improve civil rights) made up the techocratic dimension of the declaration and gave expression to this tendency.

Finally, it must be made clear that the *conformist tendency* (the *palaiokommatikoi*) was not left out of the "3rd of September" document. Although initially rather weak in the organization, this tendency, represented by the old politicians, clearly claimed most of the popular support. Their support came mainly from the middle class strata we identified in the previous chapter (i.e. farmers and the old petty bourgoisie), who made up the third and perhaps the most important factor in the societal configuration of the Greek social configuration. Thus the main attractions in PASOK's founding document for the conformist tendency were the anthropomorphic critique of foreign monopolies, along with its emphasis on the restructuring of the primary sector (e.g. the promise for the development of co-ops).

In spite of this apparently unifying nature of PASOK's appeal to the various political tendencies, the Movement could not have been so successful if it had not also been for the linchpin of a common ideological and organizational basis. "Nationalism" supplied the ideological dimension. As we mentioned earlier, the experience of the US backed dictatorship had exacerbated the already strong nationalist element in the Greek political culture. The slogan in the Declaration "Greece belongs to Greece" capitalized on just that element. This kind of slogan meant something quite different to each tendency. Interpretations ranged from an implicit critique of the traditional left, whose affiliation with the "international" pro-Moscow communist movement had worked against the struggle for socialism; to a strictly nationalist declaration of the patriotic intentions of the Movement's socialism.

The fact that PASOK was a movement and not a party served as both an important unifying function, and attracted an even wider spectrum of support for the organization. This meant that its final programme, its constitution, and the further definition of its character were to be left to the outcome of "Democratic Procedure", which was asserted to be the basis for the organization's development. At least theoretically, everything was to be laid on the table at the promised Congress, where it would be studied and decided upon. Naturally, for each tendency and for each member of PASOK this was very appealing because each was hoping to fight out its own definition of the initially vague Declaration . Therefore, it appeared that for the time being, a minimum recognition of its politics was enough. After all, the charismatic leader had confirmed that "We are not preoccupied (with anything or anybody)...we are just beginning...when the movement becomes strong, it will decide by itself (its) alliances...(and) strategic or tactical 'questions'".[36] The significance of the promise of "Democratic Procedure" will become more obvious below. It is not by accident that all the major conflicts within the party took place not over directly ideological and/or political issues, but over the organizational structure of the Movement . The structure of the decision making process as such, was the framework, or the arena in which the class struggle was internalized and fought out within the Movement.

Finally, it would be a serious omission if we did not underline the role of Andreas Papandreou in keeping such diverse tendencies together. Given the lack of any structured practice, Papandreou himself became the only guarantee that the fair play "contract" over the party's structure would be honoured. Furthermore, Andreas was unique in that he combined characteristics useful to each one of the tendencies. Among the left, Papandreou's charisma kindled the hope for making a breakthrough in recruiting the "masses", which they believed to be necessary for the further development of the Declaration's radical dimension. The technocratic tendency could identify with Papandreou's internationally recognized academic background. And finally, the *palaiokommatikoi* found in Papandreou a bridge with the past (the C.U.) and the opening for a new party arrangement which they believed was necessary for their political survival.

To conclude, the assertion that the "Declaration of the 3rd of September" was "everything and nothing", must be dismissed as simplistic and misleading. The possibility of a radical outgrowth of PASOK was in fact contained within the document and not excluded by it. The fact that PASOK did not become a revolutionary party, was not simply a result of the weakness of its founding document. It was primarily a result of the particular way in which the conflict over the document's further definition was resolved, which turned the document into "everything" for only some.

The Response

There is no doubt that despite its shortcoming, PASOK's founding document inaugurated a political discourse which overtook the traditional left. PASOK marked the first serious organized attempt to lay claim to that part of the political spectrum which had traditionally belonged to the left. The communist left, which had not yet come to terms either with its own internal crisis since the 1968 split, or with the development of the new tendencies in the international communist and radical movement, was quite disturbed. The CPs did not however, air their criticisms officially, and in fact, did not proceed to criticize the essence of PASOK's founding document. In fact, they often summed up their reservations toward PASOK by questioning Papandreou's background and motives.

Given this virtual silence on the part of the left, it is not difficult to understand why PASOK enjoyed such an explosive success among the radicalized masses, especially if we consider the early call for elections, which did not allow much time for serious organizing. Hundreds of activists and ordinary citizens responded to Papandreou's call for self-organization. They signed petitions of support, started to organize local clubs and began to prepare for the election. This phenomenon was unprecedented in the history of the Greek party system. Possibly only the organizing during the 1940s against the Germans could be seen as comparable, although obviously the conditions of occupied Greece were very different.

What was striking in the numerous petitions of support for PASOK was the social diversity of the first recruits – from contractors to workers in the building trades; from farmers to engineers; from artisans and workers, to merchants and small industrialists – they all rallied behind the new Movement. Their petitions of support displayed three main characteristics. First, there was a strong emphasis on the role of Papandreou. In most of these documents he was seen as the personified expression of the stated principles of the Movement, since he was conceived as being "the sole carrier of the call for the mobilizing of the genuinely progressive and democratic forces..."[37] Second, these petitions/announcements displayed a clear emphasis upon the nationalist dimension of the "Declaration of Principles", rather than upon the socialist dimension. They saw the Movement as an attempt to free Greece from foreign control "...to create a democracy dedicated to the service of the nation".[38] Finally, the petitions displayed a third characteristic which was similar to the attitudes of the activists who first joined PASOK. This was the belief that the Movement was genuinely new and had no relation to the corrupt parties of the past. They often explicitly rejected clientelistic practices[39] and reaffirmed their strong support for "a new organization which will be characterized by free democratic expression from the base, and which will control the leadership in political decision making and therefore obtain consistency and continuity".[40]

This eclecticism of PASOK's first recruits, despite the apparent contradictions (e.g. their overemphasis on the role of Papandreou and at the same time their strong support for the notion of "democratic procedure"), gave the Movement's dynamic a certain twist. It did not of course quell the hope of the various tendencies which had created PASOK, for the realization of their visions. However, it is only fair to say that it increased the possibility of success for some. For example, there is no doubt that the tendency to stress Papandreou's role, in addition to the nationalist dimension of the movement's first document, gave a significant boost to the *palaiokommatikoi*, as well as to the ambitions of the leadership.

Democratic Defence Joins In

While the Papandreou team was preparing PASOK's principal declaration at "Kastri", "Democratic Defence" was debating its political future. At a congress held in Athens, the organization (or, that is, those members who had not joined Karamanlis' "Government of National Unity"), developed a rich problematic around the understanding of the "Greek problem" in connection with the democratic socialist project. The final decisions which emerged in the form of a programme were quite similar to the ones finally advocated in the 3rd of September document.[41] The only differences one could detect were in the language, the strong emphasis on democracy in and outside the organization, and the less hostile and less reductionist attitude toward Western Europe. D.D.'s document was considerably more coherent and avoided unnecessary rhetoric. Its conception of "national liberation" also differed, insofar as D.D. based its understanding more on domestic class forces. For this reason both its "nationalism" and its perspective of Western Europe and the EEC were less simplistic than PASOK's appeared to be.

When one examines the documents of this D.D. congress,[42] it seems clear that it was these similarities, the lack of any real relation to the social movements and organizations and the pressure of the upcoming elections which pushed D.D. into negotiation with PASOK and eventually into joining the Movement. As D.D. recognized, it was a move of "political realism which however does not cease to be a necessary way out (based) in fact on short term criteria".[43] The organization believed that there was "no such thing as a centre-left ideology", however it recognized that there was definitely a centre-left space in the political spectrum which needed to be filled. Furthermore, it had come to the realization that "the personality of Andreas Papandreou was a powerful pole of attraction for the popular masses ... and that despite Andreas' well-known vices he is a National asset for the Movement."![44]

With this problematic on the political conjuncture of the country – one which incidently stimulated the charismatic image of Papandreou – D.D. proceeded into negotiations with PASOK. On 10 October 1974 the two groups reached an agreement in which PAK and D.D. were recognized as co-founders and equal partners in the Panhellenic Socialist Movement. The "Agreement", which was not published in total until 1983, guaranteed a certain percentage of the seats in the newly formed Provisional Central committee for members of D.D. (20 out of 75). It further gave complete powers to the Central Committee until the first congress. Every collective body of the Movement, including the prospective elected parliamentary caucus, was put under its control.[45] It recognized the need for and the significance of the "General Secretary of the Movement" (not a president) who would be under the control of the Central Committee. At this point we must note that Papandreou was the unspoken candidate for the position, although he was never mentioned by name.[46]

Finally another important innovation of this agreement, *vis-a-vis* the already published founding document of PASOK, was its recognition of organized tendencies within the Movement. "The existence and the functioning of ideological tendencies...is accepted and desirable...as long the creation of particular organized and decision making structures...(are not) formed on a personified basis".[47] In brief D.D. tried to squeeze as much as possible into the agreement so that its reservations and fears would not become realized. However, as we will demonstrate below, the power struggle over the party's policies and structure was not to be played out over the content and/or the interpretation of a private agreement. This is never the case. It was simply the naive, legalistic notion of the majority of D.D.'s members who seemed at least by implication to believe so. As we will see in the next chapter, by the time D.D. and the people within the Movement who shared their views realized the hopelessness of their struggle to control the party, it was already too late.

In conclusion, we have seen the first reactions to PASOK's "Declaration of Principles". The numerous individuals and groups who had responded positively by joining the Movement became the cornerstone of the edifice of the organization which was to "make history". However, at least at the beginning, this edifice resembled Babel more than anything else. There was a serious problem of communication and a lack of a common language which was striking. The conflict over how members should address one other – whether to use "comrade" or "sir" – is a case in point. The resolution of this tension by the adoption of "comrade" inside the organization and "sir" in public was another indication of the perspective and dynamic of the infant Movement.

Finally, we have to say that it was not only the internal choices made by the early membership of PASOK or the vagueness of its principal document which shaped its dynamic. Two more factors of an objective nature, were at least as crucial. The first was the failure of the other left political parties to challenge PASOK's base of recruiting either theoretically or in practice, and the second, the nature of the first political project which the freshly recruited masses had to undertake: i.e. the elections. The project of the electoral campaign that these recruits were to undertake was in most cases to be the first political experience recorded on the clean slate of their vague and often confused consciousness. In addition, some individuals and groups of leftist orientation did not take the election seriously, insofar as they ignored it by refusing to play the electoral game. This caused them to lose ground and strength and to suffer a strategic defeat in the power struggle over the party, with a result which was for them, catastrophic. Furthermore, all of these factors worked to the advantage of the *palaiokommatikoi* and the leadership.

Before we turn to an examination of how the first political project of the newly formed movement was conducted by its diverse membership, it will be necessary to make a brief reference to the entire political environment in which it was framed.

The Political Climate

The revelation of the US Secret Service connection with the regime, the colonels' treacherous role, the "tragedy of Cyprus" and the Junta's bloody confrontation with the students at the Polytechnic School uprising, all had a radicalizing effect on the political attitudes of the population. The most striking expressions of these attitudes were a strong anti-Americanism and a hatred for everything which was associated with the dictatorship.[48]

To characterize this attitude as permanent would be an exaggeration and indicative of a static understanding of political phenomena. There is little doubt however, that the anti-US, anti-NATO feeling encompassed almost all elements of the political spectrum. The Karamanlis government's tactical move to withdraw Greece's membership from the military wing of NATO (14 August 1974) is a strong case in point. A closer look however, at the components of this political attitude, indicates at least in retrospect that its radical dynamism, was very limited. Hatred for the USA and NATO was expressed in strong nationalist terms based on "national unity", a "unity" which started from the "government of national unity" down to the encouragement of "unity of all productive classes (in the name) of progress". The fact that national unity was at the basis of the anti-US sentiment was the main reason that the attitude never managed to be articulated in a truly anti-imperialist form. The anti-NATO, anti-USA feelings centered around the "realization" that the "allies had betrayed us".[49] This type of analysis was widespread and created the false impression that the country was somehow in equal partnership with both the USA and Britain, when in spite of the long term tradition of friendship and as fellow members of NATO, the organization had not managed to impose its principles of mutual respect among its members (Greece and Turkey).[50]

This context of heightened nationalism encouraged the identification of the "Turk" as the prime external enemy, an enemy *sui generis*, an historical enemy who is characterized by an "uncivilized barbarism".[51] This chauvinist and racist understanding of the prime enemy is ahistorical and therefore incapable of ever developing into a genuine anti-imperialist analysis, directed toward working-class based mobilization.

Furthermore, the confrontation of this enemy was seen as depending on a number of requirements: the unity of the armed forces; unity between the people and the army; and finally the maturity and responsibility of the population. These in turn further reduced the possibility of giving a radical dynamic to the anti-USA attitudes. Unity of the armed forces required that a distinction be made between the Junta and the military structure of the country. As a result of this, the dictatorship was understood, confronted and prosecuted in a personalized way. No political party dared "to identify the Armed Forces with the Junta",[52] nor to attribute the dictatorship to the army's structure or to the army's links with the USA and NATO. PASOK however, might be considered the only exception here, as it did criticize the army's structure, although admittedly in an inconsistent manner.[53] Thus overall, by implication, the "incident" of the dictatorship was to be understood as an exception and as outside the army's structure and logic. The army was to be considered neutral in the political process. Its prime function was (and supposedly had been) to guarantee national independence by protecting the nation from external enemy(ies) – (an) enemy which was already identified as the "Turk".

For this reason, the "unity between the people and army", which had been disturbed for seven years had to be re-established. After all, as the discourse of the time held, the army is nothing but the "armed hand of the nation"; it is made up of the "nation's children", since most of the officers come straight out of the "people's heart": the peasantry. Thus, the army was once again re-established as a depoliticized and neutral force in the social and political process. The communist left, in both its Eurocommunist/renaissance and pro-Moscow wings was a major contributor to the establishment of this type of discourse. In fact, for the left, this unity was articulated not simply as a desire but as something which "we should struggle for".[54]

Finally, but not less importantly, the whole theme of unity was accompanied by much advice, many guidelines and orders on how this unity must be achieved. Thus, the people were advised and encouraged to act with prudence and responsibility and were even patted on the back for doing so. References to a "healthy climate", "peace", "maturity", "national pride", etc. were common in the discourse of all the political parties. This encouragement of prudence and maturity under the widespread climate of national emergency and exceptional circumstances had an effect on popular demands. The accumulated problems and frustrations of the people came to be seen as secondary and were to be expressed in "prudent and self-controlled" forms. It was something which naturally led to a reversing and toning down of demands of the popular factor. The C.P.s once again were a major, if not the major factor in this process of castration.

Two key events highlighted the trend toward modifying the radical expression of anti-imperialist popular sentiments. These were the organizing of the first anti-imperialist demonstration in Athens (1 October 1974) and the celebration of the first anniversary of the Polytechnic School uprising (17 November 1974). In the first case the government argued that the demonstration would interrupt the "social and economic life of the city" and that it would display a lack of maturity and responsibility on the part of participants. It was an argument that was adopted both implicitly (by the C.P.) and explicitly (by the C.P./interior – and E.D.A.). Another case in point of the Movement's fluid condition in these first steps of its development was the participation of PASOK's youth in this mobilization which was dominated by ultra leftist organizations.

In the case of Polytechnic School celebration, the government and all the major political parties decided to celebrate one week after 17 November so that the electoral process would not be interrupted. Obviously the right was making a clear effort to identify the goals of the uprising (clearly anti-imperialist) with parliamentary democracy. Once again the demonstration went on with the PASOK officially leaving the decision on participation to "the consciousness of its individual members". Participation in these two events – not insignificant but not as massive as at the official events – indicated not only the trend of anti-US, anti-NATO sentiments towards moderation but also that there was clearly a struggle for the radical definition of these demands. It was also clear that PASOK alone among the left parties, though not fully nor consistently, came closer to defining these demands in a radical way.

As we have mentioned, the other important ideological result of the dictatorship experience was a widespread hatred of the Junta and everything(one) associated with it. The dictatorship was juxtaposed against an abstract notion of "democracy", which was idealized to such an extent that it was seen as the answer to all problems. Since the dictatorship was seen through the emotional eyes of the victims as a "momentary crime", as the "irrational" activity of a small number of officers, "democracy" was conceived more as its natural opposite. Democracy was the natural regime with the dictatorship as its abnormal counterpart.

This "manachean" discourse and the mentality,[55] not only presented the dictatorship in apolitical colours but also limited the possibility of giving a radical meaning to the notion of democracy itself. Thus democracy was promoted not only as a natural condition, but also as the requirement "for the nation's rebirth". This was a rebirth which was to be achieved through the "moderation of political passions" and through good rational/"democratic" management of public affairs; and since all the economic evils of the country had been attributed to the Junta's mismanagement (as we have seen in the previous chapter), democracy could guarantee economic prosperity as well. The key notion in this whole definition of democracy was "national unity and political moderation". Of course the right and centre parties have a traditional tendency to give this dimension to democracy; the left, however, in its natural role as opposition can in effect be more successful in this moderate definition of democracy. The Greek left, at times with the exception of PASOK, was a very good example of this. Either with explicit statements and slogans such as "Karamanlis or the tanks" (M. Theodorakis) and/or through implicit acceptance of the Right's definition of democracy as shown in the two anti-imperialist demonstrations mentioned above, the left helped a great deal to channel the *metapolitefse*'s strong but inchoate pro-democratic tendencies into moderate expressions. The latter in fact has been identified as the basis of the right wing's success in concealing the connection between the Junta's regime and the one prior to 1967 which it dominated. In fact this success along with the renewal of right wing personnel resulted in the unprecedented victory of the principal right wing party (New Democracy) in the November election.[56]

All the major political parties in the immediate post-1974 period participated in the dominant political discourse described above. It was a political discourse which did not faithfully express the social relations and therefore cannot simply be reduced to them. However, with its explicit or implicit statements and practices, as well as with its silences, it reconstructed the shaken hegemonic conceptions and definitions of politics. The left, which accepted the dominant discourse in a tactical compromise, did not put up a fight and subsequently was almost neutralized and led to withdraw from the construction of an alternative strategy. This internalization of the dominant political discourse by the left contributed significantly to the smooth transition from military to civilian rule. This is the only way to explain how an "overpoliticized" electorate brought "New Democracy", the party of the politician (K. Karamanlis) who had been completely discredited just ten years before, into power with an unprecedented majority.

PASOK however, was a peculiar exception to this co-optation into the mainstream political discourse. The Movement abstained from this race of compromises on the part of the other parties. At least as it appeared in its opening statement and despite its vagueness and contradictions, its discourse cannot be blamed for the creation of the moderate political climate. In fact PASOK's discourse could be considered the only "anti-logos" among the left parties. But it was only that, an "anti-logos", which was full of voluntaristic and utopian elements. It was a dynamic discourse but not a concrete one. To counterpose a developing definition of politics is not enough. This can only be the beginning. The "anti-logos" must become "logos", the antithesis the thesis and the opposition the concrete alternative. To be successful in a such a process you need a vehicle, a political organization in which these will develop. You need an organization in which the utopian and voluntaristic elements working together will develop policy alternatives based on a different definition of politics. You need an organization which is necessary for the creation of political self-confidence in its membership. This self-confidence is necessary in order to consolidate the initially instinctual political choice of the membership and in turn prevent that membership from drifting away.

All these were elements in PASOK's discourse. However, its failure to impose its own discourse or rather to develop and promote an alternative one, must be attributed not only to the social and political diversity of its membership, but also to the particular political conjuncture. Obviously the sudden election was not the best environment for the radical part of its membership to move from the politics of protest to the politics for power. Such a transition would have to be longer term, especially if the promise of democratic procedure was to be kept. The rhetorical speeches delivered from balconies by the leader created a distance which in effect turned the supporters into mere receivers. This type of electoral activity helps to inhibit the masses' participation in, and reaction to, the political drama, since it makes the leader more removed from, and less accountable to, the masses. Thanks to the election and to its charismatic leader, PASOK became a prime example of this pattern.

In spite of the unfinished and primitive nature of the *metapolitefse* political discourse, the predominance of the left in the resistance did lead to the creation of an unprecedented political atmospere by the second half of 1974. It was unlike any that had existed in the past, and was much more tolerant of radical ideas. Open anti-communism became illegitimate, supported only by a few pockets of fascists. Under these conditions, the exchange of radical ideas, through journals and newspapers as well as at university amphitheaters became an every day phenomenon. Radical political book sales boomed and cultural events took the form of mass demonstrations.[57] However, despite all this, the politicization of the *metapolitefse* was not a mass one. The great majority of the people, the "masses" that the rhetoric often made reference to, were on the whole, passive. They followed the unfolding spectacle of "radicalism" not thoughtfully but rather like football fans, whose team was somehow on a winning streak. In restropect, it was a phenomenon which was created mainly by a number of independent and left wing activists who were involved in the resistance and were trying to readjust their politics and claim some territory in the unsettled political scene.

Beyond the Maoist groups which had existed actively during the dictatorship, a number of attempts were made to organize the independent left. These attempts took forms ranging from petitions to conferences of several weeks in duration. The participants were mainly activists from the old left (E.D.A., C.P.s) and most of them had been involved in the resistance. All of these groups voiced strong criticisms of the organizing patterns and ideological dependence of the traditional left.[58] However, the rapid political developments, the diversity of the individuals involved and the pressure of the upcoming election hindered these attempts. Furthermore, the creation of PASOK, which encompassed many of the concerns of these small groups; and its subsequent recruitment of many activists, essentially eliminated the possibility of synthesis in this part of the left spectrum.

In spite of this ineffectiveness of the left, the limited effect of PASOK's discourse and the circumscription of the post-1974 radical political discourse, a number of changes did occur. There were not only policy changes but more importantly, changes occured in the attitudes of the power bloc. There were strong indications that the state and the power bloc wanted to move away from the old patterns of oppression and exclusion of "guided democracy" and closer to a more modernized one. The economic/hegemonic crisis of the early 1960s, as we saw in the previous chapter, and the experience of the dictatorship were forcing the state to undertake a number of reforms. These reforms were intended on the one hand to assist the dominant classes in smoothly resolving their differences on accumulation patterns and on the other hand, to incorporate, rather than to exclude the opposition.

When actually introduced by the government of "National Unity" and the subsequent government of "New Democracy", these reforms were more than sufficient to create a climate of "modernism" and even radicalism in the Greek context. Examples of these reforms include: Greece's withdrawal from the military wing of NATO (which lasted for four years); the legalization of the C.P.s; the punishment (although limited) of certain members of the Junta; the recruitment of members of the left into the state apparatus (although never into high positions); the democratic/unbiased elections and the referendum which resulted in the ousting of the Monarchy; the significant relaxation of policing measures; gradual language reforms; the introduction of an unprecedentedly liberal constitution; and finally the tacit acceptance of the trade union and workers' struggle as part of the democratic functioning of the system.

The First Electoral Campaign

The election of 17 November 1974 can, without great exaggeration, be called the most democratically organized election in modern Greek history up to that point. Under the conditions just described, PASOK was called upon to run its first electoral campaign. It was this election which not only delineated the various currents within the Movement, but was also to have a radical effect on its subsequent development.

The seventy-five member Provisional Central Committee of the Movement [59] put the lists of PASOK's candidates together. They did so with the advice of some old politicians who had considerable experience in electoral campaigns and knew the particularities of most constituencies. One striking characteristic of these lists stands out. They were dominated by old politicians or at best, by people who were members of PAK of the interior or associated with the "centre-left" of the mid-1960s. D.D. members along with other independent leftists and the majority of the PAK activists – of course with some exceptions – did not participate directly in the electoral campaign. They remained in the organization and their participation was "filtered" through the party structure. The underlying assumption on their part was to try to strengthen the progress of "self-organization" and through the collective bodies of the Movement to keep the candidates under control. This strategy appeared plausible at the time since even the general statement of principles put the parliamentary wing of the Movement under the Central Committee's jurisdiction. Furthermore, the call for "self-organization" was receiving tremendous response and the prospect of developing a strong organization looked very positive. Thus, the majority of the functionaries in the national headquarters of the Movement sat in the back seat during the electoral race and assumed that by putting forward guidelines they could have an impact on it.

However, there is something unique about electoral campaigns and parliamentary competition in general, which does not leave the principles of the party structure untouched. This was particularly so in the case of PASOK, since it was lacking an established practice. PASOK, being in the process of developing its own character, was more susceptible to the "magic" of electoral competition. Thus, the appearance of PASOK in the electoral campaign was not that expected by the party activists. It ran a contradictory two-faced electoral campaign. It was a campaign, which not only exposed the diversity inherent in the Movement at its outset but also the naivete and the political inexperience of the "left-wingers" in it.

On the one hand, the functionaries, relying on the local clubs and on the youth, tried to infuse a radical appearance and rhetoric into the Movement's electoral campaign. In their guidelines, they favoured an emphasis on PASOK's program rather than on individual candidates' images. The list of slogans given to the local and vocational clubs by KE.ME.DIA., the "Centre of Studies and Enlightenment" (the theoretical bureau of the organization), which was mainly run by Trotskyist and PAK (of the exterior) elements, bordered on ultra-leftism. Obviously the diffused voluntarism and the vagueness of the Movement's discourse had created room for such initiatives. Slogans such as "Down with Exploitation", "Get Foreign Capital Out", "Workers – Peasants and Students", "Forward to Build Socialism", "The Capitalist State Exploits You", and the most controversial, "Socialism on the 18th" (the day after the election) highlight the tendency of a good part of the organized Movement *vis-a-vis* the elections.[60] Very few candidates adopted this kind of discourse in their personal campaigns. Only the candidates who were closer to the organization and further removed from the practices of the *palaiokommatikoi* were paraphrasing the "3rd of September Declaration" in their personal literature and emphasizing the principle of "democratic procedure" in the building of the new Movement. Finally, these candidates made an effort to connect their personal history with the resistance, which in any event was done by the entire political spectrum.

On the other hand, the majority of PASOK's candidates ran campaigns along very traditional lines, both in the countryside and the urban centers. Their campaigns were not only removed from the local clubs of the Movement but were also isolated in a combatative way from their fellow candidates in the multiseat ridings. In addition, these candidates made an explicit effort (with some variations), to associate themselves with Papandreou, and with the tradition of the *Anendotos* struggle of his father. Meanwhile, they were trying to downplay PASOK's program, avoiding the use of the word socialism, while appealing to some vague notions of democracy and a "rational" (non right-wing) running of the country.[61]

Between these two contradictory and often controversial electoral campaign practices stood the campaign of Andreas Papandreou. Papandreou's campaign can be seen as a third expression of PASOK's diverse orientation in its first major political project. His rhetoric and appearance (he was the only candidate who wore a turtle-neck and not a tie) resembled to a somewhat greater extent, the politics of the "radicals" but he made no effort to criticize or even to differentiate himself from the *palaiokommatikoi* practices. Criss-crossing the country, Papandreou managed to draw and to move large crowds especially among the youth and the young white-collar professionals. His charismatic qualities and the fact that he avoided aligning himself with any of the Movement's currents, consolidated him as "PASOK's most valuable asset" and established him as the ultimate referee in the Movement's development.

The Result

The vitality of PASOK's electoral rallies and the apparently positive response to the Movement were however, not translated into votes. Although admirable in its first electoral attempt and only seventy five days from its creation, the Movement won only 13.58 percent of the popular vote and 13 seats claiming only third place in the three hundred seat Parliament.[62] Although under different circumstances this result would have been considered a victory, the majority of the Movement – given the political space created after the Junta – was shocked.[63]

The election was in fact a landslide victory for Karamanlis' New Democracy, which won 54.37% of the popular vote and an unprecedented 219 seats. The real loser was the Center. Although under a new name, Union Center – Democratic Forces, and despite the inclusion of some popular social democrats from the resistance, it fell drastically short of its pre-dictatorship popularity with 20.52% of the popular vote and 60 seats. It did become the leading opposition party, but it did seem that the writing was on the wall and that its fortunes were on the decline. Finally, the United Left, an opportunistic electoral schema put together by the two major adversaries of the traditional left (i.e. the C.P. and the C.P.Interior) won only 9.45% of the popular vote and only 8 seats. This was rather a poor result for the two C.P. s, even if you consider that they had been banned since the civil war.[64]

There are unfortunately no data available concerning the social or class distribution of the vote, but, judging from the regional distribution, we can say that PASOK's support was rather equally distributed between the countryside and the urban centres. In the urban centres of Athens and Salonika, for example, the Movement won 12.5% on average, which was slightly less than its national average. However, we should not be misled and conclude that the countryside voted consistently more for PASOK than the cities. In fact, the Movement's vote in most of the ridings in the countryside was much lower that in the cities. PASOK did do extremely well in the regions with long anti-right, liberal traditions, such as Crete with 21.7%, Dodecanese with 21% and Achaia with 24.81%. It was areas such as these that brought the countryside's contribution to PASOK's vote to a level above that in the urban centers. Overall however, we cannot really establish a clear pattern with regard to the regional distribution of PASOK's vote. Only one thing is certain, that "young" Papandreou's party seemed to have done very well in the same areas as had his father's Center Union in the elections ten years earlier.[65]

The controversial image of the Movement during the campaign, however, cannot be seen by any means as the main reason for the result. After all, as we have seen, the contradictory behaviour of the Movement can be understood only as an expression of the diffuse and unrefined current in the discourse among the population at large. The fact that PASOK did not do as well as expected can be attributed to a number of factors. First, the right had an amazing capacity to alter the radical post-Junta popular sentiments. The left at the same time seemed incapable of putting forward their definition of the diffused radicalism. Second, there was a legalistic exclusion of most of the young vote,[66] which was alleged radical and had rallied behind the Movement. Third, there was a real lack of mass media support for PASOK.[67] Finally PASOK suffered from limited financial resources. Its finances were based, as they had been during the resistance period, on individual contributions of the membership which given its small size could not amount to much.[68] Even if we consider the larger contributions in money and in kind of various progressive small-businessmen, PASOK's financial resources were rather limited, particularly relative to the almost unlimited depository of the right and to the regular inflow of assistance available to the established left.

In spite of its relatively "poor" performance, the elections did have a radical impact on PASOK's subsequent development. Insofar as the electoral process stimulated Papandreou's role by making him the exclusive spokesman of the organization, the political discourse was further removed from the jurisdiction of the rank and file. Such speech delivery by the leader/orator from the balconies in open rallies creates a hierarchy and a distance between the "transmitter" and the "receiver" fostering a passivity on the part of the latter. It is normally a passivity which not only frees the orator from any control but also neutralizes and incorporates any response from the masses, especially when there are (as in the case of PASOK) no established patterns of organization capable of altering this unilateral creation of discourse.

In addition, the understanding of the electoral result as a failure by both the *palaiokommatikoi* and the anxious voluntarist left activists (mainly PAK exterior and Trotskyists) created an atmosphere anything but conducive to the Movement's promise of democratic structure which was to facilitate the "equal participation of the members in the formation of the program and organizational schema..."[69] After the election, both the *palaiokommatikoi* and part of the "left" (especially the one closer to Papandreou), although obviously for different reasons, showed very little interest in trying to establish the participatory democratic structure still sought by many elements within the Movement.

Thus, it is not surprising that PASOK's main pre-occupation after the election was the organizational question of the Movement. As we will see, it was a process shadowed and framed by the after-effects of the election. It will be important to keep these effects in mind if we are to understand the subsequent disruption in the balance of power among the various tendencies within the Movement and its predictable outcome.

Notes

1. *Rendez-vous me ten Istoria* (Rendez-vous with History) Athens: Kaktos, 1981. p. 241.

2. See: "The Necessity to Modernize the Political Parties" presented by A. Papandreou in the *Greek Heritage Symposium* (6 October 1965) and published by C.U. in Italy in June 1967.

3. Interview with D. Vasileiades 17 July 1984.

4. A. Papandreou's letter to the Secretary General of C.U. and E.D.E.N., S. Kaikes (Ottawa, 10 October 1971) published in D. Vasileiades, *PAK – PASOK: Mythos Kai Pragmatikotita* (PAK-PASOK: Myth and Reality). Athens: Dialogos, 1977. pp. 118-120.

5. Both in the context of this chapter, and of the whole study, I have considered (defined) political discourse as the totality of tacit, explicit and/or symbolic actions and speeches which compose the parameters of a certain relationship. In earlier capitalist social formations, it was possible to distinguish more than one discourse i.e. private and public, since the private and public spheres were allegedly separate. Under the conditions of late capitalism, where "the personal has become political" we would call political discourse the totality of the tacit, explicit and symbolic actions and rhetoric, which surrounds and conditions all human activity.

6. *Agonas* (Struggle) 28/7 and 27/9, 1973.

7. Some of the members of PAK(interior) went so far as to refuse to sign petitions with the communists in the Junta's prisons: Interview with A. Polites, member of D.D., July 26, 1983.

8. In Germany, PAK had smooth relations with the C.P., while in Bologna, it developed good relations with the P.C.I. and "Manifesto" (in Florence, PAK appeared to be very close to the Maoists), while in Sweden, PAK's oreintation was closer to the Social-Democratic Party.

Finally, in England, where the organization was weaker, it appeared to affiliate itself with whatever choices the ever changing leadership of the group in that country was making. *Anti* No. 240 2, September 1983 p. 27; also interview with D. Vasileiades *op. cit.*

9. See for example the proposals outlined in *Oi Stochoi tou Agona tou PAK* (The Aims of PAK's Struggle), 1973. esp. pp. 7 and 17.

10. See for example the proposal made by S. Kostopoulos in the "First Ideological Seminar of PAK", where in the midst of discussion of Leninism, A. Papandreou becomes "an asset for the history of the Greek and the international socialist movement... (simply because)...he is the carrier of great socialist and patriotic ideals...": in S. Kostopoulos, *Yia Ena Patriotiko Socialistiko Kinema* (For a Patriotic Socialist Movement), Athens, Karanasis Publications, 1984. p. 87. It is worth noting that this book was first published in 1975 under the title *Yea Ena Pragmatiko Laiko Socialistiko Elleniko Kinema* (For a Genuine People's Socialist Greek Movement). For a further case in point see PAK's literature, the overwhelming bulk of which is composed of Papandreou's speeches, interviews etc. – something which became a habit within PASOK.

11. In a prophetic criticism, regarding the developments in PASOK later, from the Munich branch of "Friends of PAK", we learn that "...the patriarchal form of the organization has limited its activities to the establishment of new positions and titles for the continuous transfer of various functionaries into these positions...these changes worsened the initially bad structures of the organization (and) downgraded it into an association which collects money to support slogans...the organization (has become) a secondary association without political influence, which of course functions democratically (and) keeps itself busy but (is) without the capacity of "passing" plans and proposals...at the level of decisions on the political lines and the future of PAK D. Vasileiades, op. cit., pp. 103-116.

12. Until 1971, PAK's resistance was limited to Papandreou's speeches. Later on, and after the withdrawal of the old centrist membership, there is evidence that a small munber of PAK's members attended training sessions run by Palestinians. This however, was not to result in any major direct actions against the colonel's regime. Interviews with members of the Resistance, Summer 1983. Also in M. Nikolinakos, op. cit. pp. 66-72.

13. Interview with A. Stangos, member of D.D. and of the first Executive Bureau of PASOK, 4 July 1983.

14. To illustrate, PAK demanded that Democratic Defence (D.D.) and the Patriotic Liberation Front (P.A.M.) be considered one organization in the coordinating body of the National Council of the Resistance. Second, PAK helped to undermine the resistance with its fear of red-baiting. In fact, apparently to avoid red-baiting, PAK asked for

right wing representation in the coordinating body of the resistance. Ibid. and M. Nikolinakos, op. cit.

15. This point is rather important in explaining the subsequent behaviour of this group in the internal development of PASOK later on. For example the silence that some of the PAK members displayed during the 1975 split, when the majority of D.D. members and other leftists were expelled from PASOK, can be partially attributed to this tendency, which was expessed in the Wintenthur conference. Some middle-range functionaries of PAK never felt comfortable with the opening of the organization to other groups and personalities. From interviews with D. Vasiliades, op. cit. and D. Tzouvanos, former-member of PASOK's Central Committee, 22 August 1983.

16. A. Papandreou, Introduction to the book of S. Kostopoulos, op. cit. p. 17.

17. Democratic Defence: Constitution.

18. From interviews with prominent members of D.D. (A. Stangos, G. Notaras, A. Polites) Summer 1983. See also: *Exodus* (Exit) No. 8-11. pp. 3-8.

19. Quoted in an unpublished proposal of A. Stangos for the D.D. Conference, August 1974.

20. *Neos Kosmos* (New World), Vol. 10/67. p. 9.

21. Ibid., p. 11.

22. Due to the underground conditions within which they had to operate, the Communist Party had two polit-bureaus: one in the country and one abroad with Prague as its base. There was always some friction as to exactly where the party line was to be produced. Although this friction often took the character of conflicting personalities, the political implications for the party's development and direction were tremendous.

23. M. Nikolinakos, *Antistase kai Antipolitefse: 1967-1974* (Resistance and Opposition 1967-1974) Athens: Olkos 1975. p. 56. Also in German "Winderstand and Opposition in Griechenland".

24. Ibid.

25. *KOM.EP.* (Communist Review), Vol. 6/1970, p. 23ff. and *KOM.EP.* 10/71, p. 74ff.

26. *Neos Kosmos* (New World), No. 3-4, March-April 1974.

27. The C.P. in its international broadcasting ("Radio Truth"), condemned the uprising as adventurous. The rapid success of the movement, however, led its student organization in Greece to get actively involved.

28. From the press release of the "Co-ordinating Committee". Quoted in A. Papandreou "Greece: The meaning of the November Uprising" in *Monthly Review*, Vol. 25, No. 9, February 1974, p. 9.

29. This had been one of the Andreist slogans in 1965 and later became one of PASOK's main slogans.

30. Immediately after the *metapolitefse* D.D. called a meeting of all the resistance organizations, with the hope of creating and organizing an opposition "not against but before the Karamanlis' government". The meeting was well attended by almost all the major non right wing organizations and personalities: C.P.(Kepesis), P.A.M.-C.P.interior (Kirkos, Athanasiou), PAK (Touloupas, Koustogiorgas), D.D. (Filias, Notaras), New Forces (Tsouderou) and M. Theodorakis and A. Lentakis (leaders of the old "Lambraki Youth"). In spite the seriousness of the effort, the fact that some of the members of these organizations were already ministers as well as the question of leadership pressed by the PAK representatives condemned the negotiations to failure. Interview with G. Notaras July 7, 1984.

31. *rendez-vous me ten Istoria* (Rendez-vous with History) op. cit. p. 56.

32. When A. Papandreou arrived at "Kastri" he was confronted with a huge poster of his father and a banner calling the "old man" to "wake up" ("Old-man wake up and see us").

33. Interview with D. Vasiliades, op. cit.

34. It is rather interesting to note that the majority of these Trotskyists belonged to the Pablist tendency, which is notorious for its entrist ideas and practice.

35. Interview with D. Vasiliades, op. cit., at this point we should note that PASOK's logo, "the rising sun", was inspired by the symbol used by a small Trotskyist group.

36. Press conference: 3 September 1974 published with the Declaration of Principles by PASOK's Press Bureau. p. 26.

37. *To Vema* 22 September 1974. The identification of PASOK's first recruits with Papandreou's charisma took some extreme and often comical expressions: For example, a number of folk inspired poems were dedicated to Andreas. In them he was portrayed not only as "the leader" who was also "the protector of freedom" and the "father of PASOK itself". (From a "poem" published in a publication of the Greek immigrants in Germany).

38. *To Vema*, 22 September 1974.

39. Ibid.

40. Ibid.

41. See for example *Demokratike Amyna: Ideologikoi kai Politikoi Stochoi* (Democratic Defence: Ideological and Political Aims) Athens: September 1976 and "E Anage yia Ena Neou Typou Politiko Kinema" (The Necessity for a New Type of Political Movement) proposal by A. Stangos in the D.D.'s congress August 1974. *Mimeo* Source: Personal Archives.

42. The Documents of this Congress have not been published. However, all the documents to which I had access agreed with the point made in this paragraph. A similar view is held by A. Stangos, op. cit.

43. From the proposal of the Salonikas's section of D.D. in the same congress. p. 3.
44. Ibid.
45. "E Symphonia metaxi PAK kai D.D." (The Agreement between PAK and D.D), as co-founding members of PASOK. *Mimeo*, Article 2. Source: Personal Archives.
46. Ibid. Article 5.
47. Ibid. Article 6.
48. Roy C. Macridis, *E Ellenike Politike Sto Stavrodromi* (Greek Politics at the Crossroad) Athens: Hellenike Euroekdotike, 1984, p. 45; also in English by The Hoover Institute Publications.
49. It is rather interesting that the C.P. of the interior did not fall short in contributing to this type of analysis. In an article in their newspaper we read: "The Americans and NATO betrayed us. They knew – organized the coup and pushed their people to execute it". *Avge* 17 August 1974. See also the articles "E Hypologistes tes Atimias" (The Computer of Dishonesty). *Avge* 12 October 1974 and the reprint of Iliou's interview to "Deutche Velle" in *Avge* 25 October 1974.
50. G. Mavros, leader of the Center Union, Deputy Prime Minister and Minister of External Affairs in the government of "National Unity". Quoted in the daily *Kathemerine*, 20 September 1974.
51. Again the traditional left did not fall behind the right or the centre in this type of jargon. In an article in *Avge*, 7 September 1974 we read: "The Turk is agressive, disgusting, provocative... His role is the one of terrorist, bashibazook, of satrap.
52. It is rather predictable that the right wing or the liberals promoted this kind of analysis. The suprise is that the left with almost no differentiation promoted the same point of view; it was something which made the dominant discourse more convincing and effective. Ch. Florakis, secretary general of the C.P. on a speech on national television made it clear: "We do not identify the Armed Forces with Junta". *To Vema* 15 November 1974.
53. The emphasis of A. Papandreou on cleansing the army's structure and penalizing the army officers who were involved with Junta are case in point. See for example: A. Papandreou's speech in Salonika in *To Vema*, 3 November 1974; Idem. interview in *Athenaike*, 9 September 1974; Idem. speech at the major electoral rally in Athens, 14 November 1974.
54. From electoral speech of Ch. Florakis: *To Vema*, 25 October 1974.
55. This is more true for E.D.A. and the C.P. of the interior than it is for the C.P. See for example: H. Iliou, "Oi Ekloges" (The Elections), *Avge*, 21 September 1974; Scholiastis, "Tases dai Thetiko Atotilisma" (Tensions and Positive Outcome), *Avge*, 13 October 1974; "Anakoinose – Apophase tes Kentrikes Epitropes tou K.K.

From Resistance to Political Movement

Esoterikou" (Declaration – Decision of the Central Committee of the C.P.Interior), *Avge.* 5 September 1974.

56. J. Kaftatzoglou, *Political Logos and Ideology: The Metapolitefse Election.* Athens: Exantas, 1979. p. 35.

57. According to one source 15,000 volumes of the USSR's Academy edition of *Political Economy* were sold in the first six months of the *metapolitefse.* Some musical events at which music which had been banned for seven years was played, were attended by more than 50,000 people.

58. In addition to such groups as "77", "400", "Chaos", M. Theodorakis' "Movement of the New Left", which composed a good part of the independent left, numerous organizations and sects made a claim on the tradition of the "orthodox" left. A total of 21 "parties" claimed the title of Communist Party.

59. See appendix II.

60. From PASOK's Electoral Slogans: Internal document, *Mimeo.*

61. From candidates' literature and personal observation of their electoral campaigns. In some of these, the name of A. Papandreou was more dominant than the name of PASOK while in others the term socialism was completely absent.

62. See appendix III; PASOK's support was rather equally distributed between the urban areas and the countryside. In Athens and in Salonika, for example, it got 12.5 percent of the vote and in the countryside slightly more its national average. M. Drettakis, *Vouleftikes Ekloges 1974, 1977, 1981: Syngritike Melete* (Parliamentary Elections 1974, 1977, 1981: A Comparative Study), Athens, 1982. Ch. 2.

63. Papandreou was expecting at least the leading opposition role in the new parliament if we can judge from preparations on the night of the elections.

64. See also Appendix III. For a more detailed analysis of the vote distribution by region, see M. Drettakis, op. cit. ch. 2.

65. For example, in the same area noted here as PASOK strongholds, G. Papandreou's C.U. had managed absolutely unprecedented gains, e.g. 80.7% in Crete, 68.25% in Dodeganese, 57.8% in Achaia. *Odegos Eklogon op to 1961* (Electoral Guide Since 1961). Athens: Vergos Publications, 1977.

66. The election was run basically on the 1964 enumeration.

67. PASOK did not manage to gain anything more than a "neutral" response from the "Lambrakis" liberal press conglomerate, which is unquestionably so influential that it can vote governments in or vote out of power.

68. These contributions never exceeded the amount of 250,000 Dr. or the equivilent of $8000 at the time. D. Vasileiades, interview op. cit.

69. See Appendix I: The 3rd of September Declaration.

3

The Critical Year of 1975
The Consolidation of Papandreou's Control

Regardless of the reasons behind it, PASOK's call for self-organization in the fall of 1974 was a "revolutionary" initiative. The people were called upon to create their own political organization. The 3rd of September document was only the basis, the beginning of this process. The Movement was about to engage in a march towards its self definition, which was to crystallize in a national congress.

This idea of self-organization not only generated some important theoretical questions but it also created a number of problems in practice. At the theoretical level PASOK waffled around the questions concerning the Movement's organizational practice during the interim period until the first congress. Would the collective bodies of the organization be elected or not and if not to whom should they be accountable? Would the expression of tendencies within the Movement be allowed and if so to what extent? Would the decisions of the majority be compulsory for the minority and if so would the minority have the right to express its views within the organization? And finally, would the collective bodies of various levels be allowed to communicate with other collective bodies of the same level without the mediation of the higher level and more centralized bodies? Such were the questions which dominated these discussions after the general elections.

At the same time, as seems clear from various internal reports of the Movement, the call for self organizing had run into a number of practical problems. This was particularly true in the countryside where the experiment had to confront the old clientelistic practices. The "initiative committees" (or at least 60-70% of them) were simply made up of ambitious individuals who wanted to be nominated for M.P.".[1] According to the same reports, the people who had taken the initiative to organize the first committees of the new Movement were middle class professionals. In at least one of these reports, such phenomena were identified as the prime reason for the lack of a genuinely mass recruitment among peasants and workers, since these strata were traditionally associated with the leadership of "the establishment".[2]

These questions and problems which the party faced in the post-election debates arose as a result of PASOK's idea of self-organizing. They contributed to the problems encountered when the party made its organizational question the prime focus of its post-election activities and they were expressed in the dispute which arose around the Pre-Congress initiative. It was around this initiative that the first major contradiction, polarity, and eventual split of the newly born Movement developed. These events in turn helped to steer the Movement away from a radical revolutionary interpretation of its September 3rd promise and to set the pace for PASOK's subsequent development.

The Post-Electoral Debates

Under the impact of the electoral result, the debate over these problems created two distinct political poles within the Movement. The most diverse groups of the organization were tactically allied around these two political points of view in a peculiar manner. Although the borders of the two poles of debate were rather diffused, with no great risk we can identify: on one hand the "Leninist" tendency of PAK (which happened to predominate in that part of the organization which was outside the country and which later came to be known as "PASOK B") and the *palaiokommatikoi*; and on the other, the majority of D.D. members, and members of communist or generally leftist background.

The major questions of these post-electoral debates were centered around the Movement's constitution, the internal organizational principles of the Central Committee and the orientation of the trade union wing of the organization. However, given PASOK's primitive structure, the discussion of these questions remained the privilege of the members of its centralized collective bodies, and they never became a direct focus of the rank and file.

The majority of the written proposals came from members identified with the second of the above mentioned poles (i.e. D.D. and the independent left). According to their proposals, the building of PASOK's structure should be a long term project if it was to be based on democratically structured mass participation. With this in mind, they planned a number of surveys and educational programs which aimed at stimulating rank and file awareness and participation. They conceived of an organizational edifice based on rank and file clubs. In the same vein, they understood the role of the middle range and top collective bodies of the Movement as having coordinating and not decisive powers.

In addition to these organizational principles, there were important ideological and political implications contained in the proposals. As derives from its rather sophisticated constitutional clauses and charts, this current held a radically new, pluralist definition of politics. This pluralism referred not only to a simple confirmation of the already accepted principle of recognizing various political tendencies within the Movement, but also to the definition of the spheres of political activity. The latter was rather important since for the first time within the context of the Greek party system, equal weight was given to spheres of politics beyond the traditional framework of electoral competition, student, or trade union related issues. In most of these proposals, structural recognition was given to issues related to women, youth, culture, environment and "quality of life". Obviously there had been a great influence from the left developments in Western Europe. The functionaries saw a development of a socialist movement which would incorporate all these issues in the main corpus of its politics without reducing their significance. Thus, in their organizational proposals, there were clauses which, both allowed the separate structural expression and development of all these non traditional political concerns, and guaranteed their contribution to the formation of the "Movement's line".

On the question of the constitutional principles of the Central Committee, the problematic of these functionaries began with a critique of its practice up to that point. Although they recognized the constraints of the electoral campaign, they were very critical of the higher body of the Movement. The provisional Central Committee had not functioned as such in the first three months of the Movement and its functions were undertaken by Papandreou and the Executive Bureau. Thus, the new proposals tried to undercut the further concentration of power in the hands of Papandreou and the Executive Bureau, which had been appointed by him. They saw the provisional Central Committee as the highest coordinating body of the process leading to the Movement's first Congress. It was allegedly there that all the political, ideological and organizational problems were to be resolved in a democratic manner. According to these proposals, all the functions and activities of the Movement were to be under the supervision of the Central Committee. This was to include the entire parliamentary wing. In fact, the latter's participation in the central bodies of the organization was considered a conflict of interest.[3]

The discussion on trade unionism did not appear as such to be the focus of intra-Movement conflict. However, it seems clear from the existing internal documents that the "Trade Union Bureau" of the Movement (occupied by members of D.D. and the independent left) was promoting notions and practices which were at least implicitly foreign to both the "Leninists" and the *palaiokommatikoi*. The documents of the Bureau were very critical of the sublimation of the trade union movement in the electoral campaign. Thus, for the future they were suggesting that there must be some level of autonomy between the political organization and the mass trade union movement. That is to say, trade unionists in the Movement should never act as mere transmitting agents of the Movement's politics. In this way not only would the much desired theoretical principle of "the autonomy of the mass movement(s)" be served, but more importantly, PASOK had much to gain from the problematic and the practice of these movements.[4]

There is little doubt that the majority of the written proposals were permeated by these ideas. Of course this does not mean that the *palaiokommatikoi* and/or PAK were absent from this debate. However the nature of their ideas on how the Movement should pursue its development did not require the sophistication and the commitment which the above mentioned written proposals had entailed. Thus they merely sat back, often tacitly, and occasionally explicitly, expressing their dissatisfaction with the developing problematic in the Movement.

Though for different reasons, both the *palaiokommatikoi*

and PAK did not like the idea of structuring the Movement along the lines delineated in the written proposals. On the one hand, the *palaiokommatikoi* saw a threat to their future in these proposals. A democratically structured mass organization, which by its structure and its political activity undermines the parliamentary definition of politics, was obviously not in their interests. They would have preferred a much looser organization around a charismatic leader, geared to run successful electoral campaigns. An organization like this would obviously be more amenable to their clientelistic practices.

On the other hand, the old PAK members had strong objections to a pluralistic development of the organization in which the rank and file clubs would be in command. After the establishment of PASOK, the membership of PAK, which was based abroad, did not completely merge within the newly formed organization. According to several sources[5], they continued to function as a separate and closed group. They were identified as "PASOK B" and their strongholds were basically abroad (Italy, Switzerland, West Germany) and to a lesser extent within the country (Patra, Salonica). They considered PASOK as a front of PAK and their main goal was to promote its transformation – very much along the lines of the Leninist idea – into a closed vanguard party.[6] Their legitimacy was derived not only from the "glorious history" of PAK but more importantly their close association with Papandreou.[7]

This group's voluntarism and its self-indulgent idea of vanguardship put it objectively in a hostile position *vis-a-vis* the above mentioned proposals on the Movement's future. The vision of an organization which develops slowly but on a democratic basis, in which various political tendencies freely co-exist, an organization marching towards a congress where everything was to be tabled and decided upon, did not coincide with their political ambitions and anxiety.

In spite of the skepticism of these two groups, the call for self-organization created a very vibrant atmosphere within the Movement. It was an atmosphere unknown in the history of the Greek parties. Members of the rank and file organizations, especially those with some intellectual background, initiated numerous theoretical discussions and publications. With the assistance of the part of the Movement's leadership associated with the idea of a decentralized and a democratically run organization, those apparently more theoretically astute members of the rank and file clubs developed a debate in an attempt to further refine some of the themes of the "3rd of September Declaration".

As a result of this debate, a number of pamphlets and articles were published by various collectives and individuals. Themes such as "transition to socialism", "self-management", "the relation of social classes to the Movement", "capitalism and the education system" as well as analyses on various liberation movements, dominated these debates.[8] In addition to these theoretical questions, a considerable emphasis was given to the study of current political questions. In fact, a number of study committees were formed for this purpose. However these committees rarely managed to finish their projects. The development of friction around the question of the Movement's structure overshadowed their efforts. The only exceptions to this rule were the committee which put together PASOK's own proposal for the country's constitution and the student section of the youth, which produced a comprehensive program for the student movement.[9]

The content of these theoretical exercises was not particularly impressive. They were usually characterized by a simple mindedness, a pretentious superficiality and a naive simplicity – especially on the issues of socialist transformation. However, in the context of a country which not only lacked the tradition of a serious left intelligentsia but also had just come out of a seven year dictatorship, this discourse initiated by the first recruits of PASOK should be considered rather important. Its significance lay not only in the fact that it was permeated by left Marxist ideas in which the new left concerns were not ignored, but mostly in the fact that it hinted at the potential and the direction which some of the organized activists might take. There was at least an implicit move away from old left wing politics and an attempt, though not well articulated, to create something new.

The activity of the various study committees had significant implications for the ongoing arguments over the Movement's structure. Obviously the idea of forming policies according to theoretical analysis and/or concrete studies did not appeal to *palaiokommatikoi* nor to the "Leninists" or even to Papandreou himself. To the *palaiokommatikoi*, sophisticated theoretical discussions and written political comments, which they could not influence or even follow, would not only frighten their clientele but would also not allow them much room for political manoeuvering during elections. The "Leninists" (PAK) also did not find the content and the direction of this problematic very favourable. Although they contributed to some of the theoretical analyses, they saw the overall new political flavour deriving from these analyses as a threat to their idea of socialism. Consequently, they denounced some of these ideas (e.g. self management, the autonomy of the mass movements from the political organization) as "liberal" especially since some of them presupposed a quite open organizational structure. Finally Papandreou himself was not very keen on this growing pattern in his Movement. Without being vocal about it, he definitely did not encourage it.[10] An emphasis on systematic studies and written commitments on policies would have undermined his charismatic role in the organization and tied his hands in directing it according to the political mood of the situation. The institutionalization of this method of producing policies would not only have strengthened the collective bodies of the organization but would have implied his accountable participation in some of them. This of course was something that Papandreou's ambitious personality could not tolerate. He had avoided it throughout all his public life even, including, as we saw, during the period of the dictatorship.[11]

The friction which developed around these issues did not remain only at the level of the various proposals or abstract articles on strategy, nor was it only to be identified in the attitudes of the two poles. It soon crystallized in a compromise in the collective decisions of the organization. In a resolution of the Provisional Central Committee under the title "For the advancement of the organization of the Movement" we read that "the organizational development of the Movement is to be based (simultaneously) upon the principles of... democratic procedures and on the efficiency of the organization".[12]

It was the first time that PASOK had put forward a utility argument side by side with the highly popular promise/principle of democratic procedure. No one up to that point had been concerned with the question of efficiency in this context. At the same time of course no one would have denied the need for an effective organization. The occurence of the two together however, was a clear effort to define the abstract principle of "democratic procedure" and subsequently to give to the Movement a certain structural direction – one which under the circumstances could predictably only be centralized. At the same time of course we can see, even from this brief excerpt, that the leadership or at least the part which had the upper hand – very much like Poulantzas' concept of the power bloc in the wider society – did not abandon or deny the opposition's discourse. On the contrary, by adding to it and/or subtracting from it, they were able to incorporate and to neutralize it. This was a pattern faithfully followed by PASOK's leadership in its effort to take control of the organization.

The First Attempts at Control

Given the "radical" sentiments within the Movement, the leadership could not take on the opposition on ideological or political grounds. They had to do it by attempting to control the framework within which the debates were taking place. Ideological and political discourse cannot exist and moreover cannot be of any effect, unless it takes place in a compatible material environment, which, in the case of political parties, is none other than their organizational structure.

The leadership of PASOK and particularly Papandreou himself understood this very quickly. Thus, the first attempt to control the direction of the Movement was an effort to dilute the rising unity among the organization's top functionaries. On 28 December 1974, bypassing the weak voices of opposition, Papandreou announced the expansion of the Provisional Central Committee. Claiming that the Committee had to become more efficient, Papandreou appointed another thirty members to it. Of course, given the agreement with Democratic Defence, he tried to maintain the same ratio of members from the various groupings of the Movement without giving much space to the *palaiokommatikoi*.

In retrospect, we can say that Papandreou's criteria were not explicitly ideological. His appointments showed a mild preference for PAK members and generally for people who were closer to and who at least appeared to be more faithful to him. However, looking at the ninety-nine members of the new Provisional Central Committee – Papandreou never became part of this or any other collective body of the organization – we cannot detect any changes in the ideological orientation of the Movement. Despite Papandreou's claims otherwise, the result was merely to create a huge, inflexible and disfunctional Central Committee which further strengthened his dominance in the organization.

However, it soon became clear that the opposition was not about to disappear into the multitude of an enormous collective body. As we have seen in the previous chapter, with the entrance of D.D., and other groups and individuals of the independent left, PASOK became a meeting place for the radicalized intelligentsia and left activists. Its Central Committee was no exception. The members of Democratic Defence, who held the most comprehensive vision of the Movement's future, were joined by a number of other functionaries originating from literally all sides of the organization, including even some members of the old PAK.[13] Instead of diluting the sound of dissenting voices, Papandreou's move had brought them together.

This became clear during the first session of the Committee on 19 January 1975. In this meeting a fierce argument broke out over the selection process for the middle and lower collective bodies of the organization. The proposal suggesting appointment, which was supported by Papandreou himself, was defeated. The Committee adopted a resolution calling for a selection process based on elections, which at least by implication would stengthen the local clubs. It is interesting to note that the people who agreed with Papandreou on this depended on him for their political power and their careers. They had had no political/public background during the Junta period, which was particularly important in the Greece of 1975 in which the "glory of the resistance" carried great weight.

Confronted with the prospect of losing control of the Movement, Papandreou began to prepare his next attempt to put "things in order". In retrospect, it appears that he began his ground-work right at that meeting. In his speech delivered at this session of the Committee, he referred to "the worrisome problem...of the tendency for grouping, which has developed among the various political currents within the Movement".[14] At the time there were no organized factions in the Movement, beyond the natural sequence of old and new friendships and political bonds among activists. This does not mean that to the informed observer there were no distinct political tendencies within the party. Papandreou however, insisted that organized factions existed and that they would not only be fatal to PASOK but to the entire "popular mass movement".[15] In a skillful, machiavellian manner Papandreou was apparently trying to create the proper psychological atmosphere for the acceptance of his next move. Besides the infusion of severe psychological pressure against these factions (since political disagreement of this nature was equated with the destruction of the entire mass movement and the activists associated with it were implicated as enemies of the people), this preparation entailed the building of political support especially among the inexperienced members, who could not have fully understood the importance of the debates and could have gone either way.

Furthermore, in his effort to legitimize his "fears" of factionalism Papandreou made a rather lengthy reference to the origins of PASOK as well to a number of ideological and political issues which were being debated by the party at the time. On the ideological origin and identity of the Movement, Papandreou made no pretense of hiding the fact that PASOK was composed of "a part of the old Centre Union and (its youth) E.DE.N....another part which called or still calls itself "Andreism"...and the left in and outside the traditional schemas". This diverse origin of the new Movement was presented by Papandreou as the reason for the problem of identity in the organization. "Here, (in the political diversity of PASOK)", Papandreou continued, "and not with the 3rd of September, is where the problem lies, since everybody read into it whatever was closer to his character and psychological mood". Of course Papandreou admitted that the statement of principles was rather "general" and allowed great room for "the processing of alternative programmes". However, he was quick to point out, contrary to the general understanding up to that point, that "regardless of interpretations, not even the planned congress of the organization would have the right to depart from the 3rd of September". To him "revision of the founding statement (was) possible only with revolution". In fact he went so far as to advise those members who were hoping to change the Document to "find political refuge in another party".[16]

At the same time, Andreas made references to the ideological questions which derived from the Document. Thus, while he was advising everyone not to attempt to change the founding principles of the organization, he retained that privilege for himself, simultaneously elevating his positions as the only legitimate interpretation and consequently preserving for himself the exclusive right to put forward the operational/unifying political definitions of the new Movement. In his admirable and simplifying manner Papandreou made a convincing attempt at defining a wide range of open issues: from the nationalization and socialization plans to the questions concerning Marxism and the idea of vanguardship. What is interesting is not so much the answers to and the analysis made of these questions per se but rather the apparent intentions of the speaker. In this ten page document Papandreou skillfully managed to include the concerns and the worries of all currents of his movement and make a strong case (both implicit and explicit) for the Movement's unity.

The fact that "we are all bound by the principles of the 3rd of September must be", Papandreou argued, "the basis for the Movement's unity, and concerns such as (whether or not one) is a marxist and/or comes from the center must be ended." Andreas went on to define the ideological boundaries of the organization. On the Movement's right Papandreou saw social democracy, which at least in Europe and particularly in the case of the S.P.D. had done nothing but support the interests of big capital. Compared with these sorts of practices "we have qualitative differences", Papandreou continued, since "some aspects of the S.P.D. program have even been less progressive than those of Johnson in the US... (therefore) we are not Social Democrats".[17]

On the other hand Papandreou underlined the Movement's difference from the rest of the left. He denounced nationalizations in favour of "socializations" although he was more than usually vague on the question of the extent of these reforms: "Whether or not we are going to socialize Onassis or something more...it would depend on us, as we will be defining it through our studies and programme, as we are engaging in a dialogue with our people (and) the social classes which show interest". In fact, the only real contact with the people that took place was solely concerned with promoting Party membership on the basis of the vague founding document and the promise of democratic participation and not on actual policies, as Papandreou implied here. In addition, although Papandreou did not fail to preserve the title of vanguardship for PASOK in a rather self-indulgent manner, he made no mistake of distinguishing it from the one advocated or implied by the C.P. s. "The notion of the enlightened vanguards who lead...is quite different from ours... we do not create establishments...our vanguardship", he argued, "must be defined in last analysis, by the base". For this reason, PASOK he continued, is dedicated to democratic procedures and it is "with these democratic procedures and within the legal framework that we (will) pursue the socialist transition".[18] Papandreou concluded his criticism of the traditional left[19] by making reference to the question much debated by the organization – that of marxism.

Papandreou admitted that the "3rd of September says nothing on this question – i.e. on whether or not we are marxists – and therefore we can answer it either way"(!). "But," he continued, "if by the word marxist we mean the method of analysis which we inherited from Marx, which talks of class struggles, of the structure of power...we are obliged as a socialist movement to say YES!... For this reason, we are against the liberal notion that the state is always a neutral entity, which equally services all the citizens...(In fact) whoever believes this should drop out of politics, not only the socialist movement".[20] Having said this, Andreas was quick to point out that the marxist "dogmas", which developed in the last 50-60 years should not be the concern of PASOK. "We simply have", he added, "a method of analysis with a marxist origin".[21]

However, Papandreou knowing that he could not afford to alienate the less radical sections of his organization, decided to further qualify, or rather water down, his initial "yes" to marxism. Thus, he noted that in spite of the usefulnes of marxism, there have historically been overly enthusiastic marxists without critical judgement and maturity and therefore "the question is not whether we are marxists or not, but if we have critical judgement or not... (and) this, to him, was the greatest problem of the Movement since we have been put in danger by a narrow minded dogmatism".[22] There is little doubt that there is certain amount of truth in this argument. However, when it is presented in a milieu almost completely lacking in deeply rooted marxist debates, and understood in such an eclectic and a-historical fashion, it inevitably neutralizes the initial acceptance of this discourse and retards the possibility of its development, before even starting it.

To conclude, it seems in retrospect that Papandreou's speech was of critical importance. It laid the foundation for legitimizing the latest move in the attempt to take control of Movement's structure. With his speech Andreas not only managed to define the "3rd of September" in his own terms and preoccupy the debates of the rank and file, but more importantly he did so by not alienating the various, adverserial currents in his organization. Thus, the liberal, and more conservative elements of the Movement were able to identify with the leader's commitment to democracy and his demarcation of the traditional left, and not be frightened by the nationalization program, which was to be left to the future economic and social conditions. Nor was it necessary for them to be sceptical of the stated acceptance of marxism since the important thing was not whether "we are marxists but rather whether we have critical judgement or not". At the same time, the left wing anxiety of the movement could be satisfied through the president's leaps which asserted that "we are not social democrats", and since marxism per se was not denounced.

Papandreou's skills and ability to orchestrate such an accomplishment, elevated him more than before to the status of the "leader umbrella". This in turn, further legitimized his direct appeal for unity which in effect downplayed the existing ideological and political differences and limited the effectiveness of the debates[23] around them. The "Statement of Principles of the 3rd of September", as it had now been defined by the leader was enough to guarantee this unity. Whoever dared to challenge it "was to put him/herself outside the Movement"[24] and it became this leader's exclusive prerogative to decide who had committed this fatal sin.

The events which followed and especially the reaction or the lack thereof to Papandreou's initiatives must be understood in this context. In his attempt to take over complete control of the organization, Papandreou was to launch a direct attack against the membership, which held the potential of creating an organization in which his role would not be the dominant one.

The Pre-Congress Announced

By the end of the First Session of the Central Committee, with the passage of the resolution calling for the building of an organization based on elections and the reaffirmation of belief in the magic of the 3rd of September, the atmosphere within the Movement was calm. Only the echoes of Papandreou's speech foretold the hurricane to come. Ten days later, on 24th of January,

1975, Papandreou announced to the "Coordinating Secretariat" (the committee composed of all the chairs of the various committees), his decision to dismiss the Provisional Central Committee and call a "Pre-Congress".

Actually, Papandreou did not call for any formal, collective meeting of the Secretariat to announce his decision. Instead, he confidentially invited individual members of the Secretariat, in small groups of one to three, to his hotel suite a few blocks away from PASOK's headquarters. These invited members were among his most openly loyal supporters. Apparently Papandreou hoped to keep his decision secret for a few days so that the "Andreists" could prepare the ground for its smooth acceptance by the organization at large. However, he misjudged the attitudes of one of these functionaries (K. Manolkides) and the whole enterprise was blown wide open.[25]

It was at the least, a surprising decision since Papandreou had not only taken the initiative without consultation with any collective body within the Movement but also because the organization was anything but ready for such an undertaking. No fixed patterns of decision making had been established and therefore the organization was left open to manipulation by the various experienced functionaries who in the countryside were none other than the *Palaiokommatikoi* and the *Kommatarches*. Furthermore, the date for the "Pre-Congress" was set too early – on the 16th of March – and it was obvious that six weeks were not enough for the preparation of proposals and resolutions, let alone their discussion by the rank and file. Thus again the whole preparation process and the selection of delegates would be less than democratic. The latter consideration was particularly ironic, since an improvement in the democratic functioning of the Movement was Papandreou's main argument for this initiative.

These were the fears of most of the functionaries who initially strongly objected to this adventure. Such objections came mostly from old D.D. members, students, those who had joined the organization from the traditional left, as well as some PAK members. Although they themselves were strong advocates of democratic procedures and elected collective bodies, they felt that it was premature to initiate these processes at this time. In their opinion, the "Pre-Congress" enterprise was leading to a facade of democracy which would strengthen the views of the already strong forces in the Movement's structure, namely the

Presidency and all those engaging in clientelistic relations.

Judging from the way in which he chose to announce his initiative, as well as from the events which followed, Papandreou seems to have been aware of all the drawbacks of his decision. He apparently hoped, first through some superficial democratic procedures, to corner the membership which could potentially criticize the extent of his flexibility – their growing strength had already become clear at the first session of the Central Committee – and second, to replace them with those loyal to him. His excuse however, was that there was a developing factionalism and elitism in the organization.

Papandreou even admitted that his decision was "formally undemocratic". However his cavalier attitude is evident in one of his informal meetings with members of the Central Committee and top functionaries. The first speaker from the floor tried to convince him to stop the unfolding preparation for the Pre-Congress and instead to call for the Movement's first real Congress in the near future. Papandreou's answer to the challenge was "Comrade, you can change my opinion but not my decision. I have made the announcement to the press."[26]

His decision was justified on the basis that it was to lead to "the expression of the popular rank and file".[27] This was not enough to stop the turbulance which his initiative generated. Papandreou refused to give any more information on the extent of "factionalism and elitism" and in fact limited himself to stating that the "phenomenon is luckily not extensive"[28] and does not go beyond five to ten members of the Central Committee. These points in addition to his argument that the Central Committee was weak, were not however good enough reasons for the dismissal of the higher body of the Movement. The Executive Bureau discussed the issue, two days later (31st of January), and was divided on it.[29]

Looking back at these events, there is no doubt that Papandreou's decision to dismiss the Provisional Central Committee and call for a Pre-Congress was the beginning of the end of the hope for the building of a democratically structured socialist movement in the country. Although the decision was welcomed by a number of local clubs – especially from the countryside – the youth, the professional and great numbers of urban clubs reacted negatively. Composed of a more theoretically astute membership, these parts of the organization considered Papandreou's initiative a coup against the democratic development of the Movement. The reactions and the counter reactions, faithful reflections of PASOK's political composition, varied in intensity and sophistication, but did not fail to touch, in one way or the other, the entire membership.

On 6 February, a member of the Executive Bureau, Vasso Papandreou stormed the offices of *Agonistes*, PASOK's youth periodical, and delayed the printing of an editorial entitled "Socialism and Democratic Procedures". The article underlined the compatibility and the necessity of "democratic procedure" in the "struggle for socialism" and in spite of its generalities was implicitly, though unmistakeably, critical of Papandreou's decision. The episode, due to V. Papandreou's close association with Andreas, generated a lot of reaction especially among the rebellious youth. Finally after a long, fierce battle and with the mediation of the Executive Bureau, the article, signed by ten out of eleven members of the editorial board, was published.[30] V. Papandreou resigned.

A few days later the Youth Bureau, responding to rank and file pressure, invited Papandreou and the Executive Bureau to a common meeting in Athens, to explain the reasons which had led to the controversial decision. The meeting clarified not only the confusion of the membership, which had very little information but also the lack of even tactical unity among the functionaries of the opposition. The low tone of voices and often the complete silence of the members of the Executive Bureau who had opposed the decision allowed a lot of room for Papandreou's eloquent abilities and further isolated the unorganized and scattered criticisms of the audience. Thus Papandreou, if not a clear winner, managed to walk out of the meeting with enough consent to pursue his plans.

On 5 March, L. Philippatos, a prominent member of the Central Committee and a former functionary of E.D.A., resigned from the Movement. In his insightful and prophetic letter to the president of PASOK, Philippatos pointed out the dangers of the decision. He critized Papandreou for leading the organization into a "blind pre-congress" since the uninformed membership would have to rely on the president and his trusted friends for the election of the new Central Committee. Furthermore, Philippatos, extrapolating from his experience with the traditional left, anticipated that Papandreou's refusal to name the "conspirators of elitism" would replace the democratic procedures with a proliferation of "whispering in the organization's corridors", which would in turn split the Movement.[31]

These examples of the events which followed the announcement of the controversial decision, highlight the atmosphere in the organization as it was pushed into the pre-congress. It was an atmosphere dominated by sounds of whispering in the Movement's corridors, by backstage dealings among the supporters and opponents of the decision as well as by vicious bickering and indiscriminative slander of members of the Central Committee who found themselves on the wrong side of the issue. It was an atmosphere which did not even respect the "fighters of the resistance" as their dignity and integrity was questioned. It was amazing that such a party, which to a great extent had originated from the devastating experience of the traditional left, and a few months earlier had set off to create its democratic counterpart, could present such an image.

The Pre-Congress

The controversy concerning the decision to procede with a pre-congress dominated the preparation period. In fact, the debates in the local and professional clubs during this period did not focus on proposals of what was to be discussed in the pre-congress – in any event these were produced only a few days before the event – but rather around the question of the correctness or the incorrectness of the decision itself. Inevitably, sophisticated political arguments were put aside and the brief debates which were to lead to the selection of delegates were polarized between "Andreists" and those who opted for a more collective organizing of the Movement. The only exception to this pattern was the youth and to a lesser degree the professional clubs, which at least on the surface had the pretense of being more theoretically astute. Of course these sections of the organization also debated the timing and not the concrete proposals of the pre-congress but they did so in the context of some indirectly related theoretical issues.[32]

On 7 March, nine days before the pre-congress, the "Preparation Committee" which was appointed by Papandreou,[33] circulated a memo with the guidelines for the preparation and the actual process of the big event. From this memo PASOK's members learned that the pre-congress was to last only one day!

The Critical Year of 1975

They were also informed that there were to be three main proposals: a) "On the prospects and the problematic of the Movement" by A. Papandreou; b) "PASOK in the Greek socioeconomic space" by V. Filias;[34] and c) "The basic organizational principles of PASOK" by A. Tsochatzopoulos. The memo promised that these proposals were to arrive at the local and professional clubs within the next three days (i.e. six days before the pre-congress) and since interventions from the floor were not allowed, the delegates could bring their comments to the "Preparatory Committee" two days before the pre-congress. The comments were not to exceed one type-written page! It was becoming clear, even to the most politically inexperienced, that the strategy of the President and his close associates was to control the process of the pre-congress. In this way they could also control the outcome of the conference, which was in fact nothing other than the election of the new Central Committee.

The most interesting part of this memo was the recognition it gave to the parliamentary caucus which up to this point had been consistently undermined. For the first time, in its official documents the organization gave as much importance to its parliamentary wing as to the other sections of the Movement. The "chair" of the pre-congress was to be filled by the appointment of the "Preparatory Committee", which in turn was to be composed of "one representative of the Parliamentary Caucus, the Executive (bureau), the Organizing office, the Trade Union bureau and the youth". It was a move that highlighted the change that the Movement started to make at the time. Many[35] have claimed that Papandreou was getting a lot of pressure from PASOK's parliamentarians who represented the most conformist part of the organization. They wanted to be included in the Movement's internal processes and the time appeared right since Papandreou needed all the help he could get in putting the organization under his control. However the "entrance" of the parliamentarians into the Movement was not a tactical move. Along with their questionable cargo of *kommatarches* and *palaiokommatikoi* practices, the parliamentarians were there to stay and would soon expand their presence and influence.

Obviously, this kind of designated procedure combined with the limited time allowed, could do nothing but discourage comments and counter proposals. Thus, the fact that there were not a great number of rank and file proposals was not surprising. The ones which dared to appear came from those rank and file clubs which had some connection with headquarters – and therefore access to information. Still however, these proposals could not claim to be counter-resolutions. They were usually brief protests against the president's[36] decision and the lack of an adequate explanation for it. They protested the undemocratic procedure followed which did not allow enough time for "debate in the rank and file so that the elected delegates would be the couriers of the positions of the membership... (instead) they were asked to respond to seven page long proposals within one to two days".[37]

There were however, a couple of exceptions to these overall low key and non-political reactions to the whole process of the pre-congress. One came from the four members[38] of the Executive Bureau who had disagreed with the idea of the pre-congress from the beginning, while the other came from a more anonymous source; it was signed "Committee of Unity of Members of PASOK". The political depth and implications of their analyses and the reasonably wide-spread respect they enjoyed made their documents the focus of a fierce and unfair attack/response from those members of the Executive loyal to Papandreou. Although these proposals were to be discussed as alternative resolutions by the delegates at the pre-congress, the Executive which was in effect composed of the president and his close associates, circulated a polemical disclaimer of the proposed resolutions within the organization. The "Executive" called these proposals divisive, and undisciplined, and characterized the members associated with them as lacking comradeship and ethics.[39]

If one reads the exchanges between the two sides on the eve of the pre-congress he/she might wonder how it was possible for the Movement to remain united. Members of the organization fought and stabbed one another in the back not necessarily on clearly defined political issues. Some members even questioned the moral integrity of others – completely outside the constraints of accepted organizational patterns of communication (for example the use of slander, etc.). At such a point, the question of unity becomes rather intriguing. In the case under examination however, the phenomenon is hardly inexplicable. Something in common to both sides of the controversy allowed the organization to march into its first major collective process at least temporarily united. This was on the one hand, the stated promise that the Movement should soon have its first congress proper, and on the other, the hope of the opposition for the realization of this promise. Allegedly it was there that all the differences would be tabled and fairly decided upon.

In spite of this unifying point, the despicable atmosphere within the organization during the previous six weeks was in fact transferred to the actual proceedings of the pre-congress. The luxury of the conference room in the downtown Athens hotel could not diffuse the intense Middle-Eastern bazaar-like atmosphere. The Movement was so divided that an uninformed observer could easily have concluded that he/she was dealing with the conference of two different, hostile political organizations. Many delegates were not even talking to each other. The "President's men" were selectively circulating a list with the names of the delegates Papandreou "wanted to see in his new Central Committee". The uninformed and/or misinformed[40] rank and file who followed the procedure as security clerks or as observers (mostly high-school and university students), acted as police in the interests of the leadership. While they were applauding the leadership, they never failed to boo and even physically attack those attempting any form of criticism,[41] branding them "right-wing social democrats". It was an atmosphere very much reminiscent of the gatherings of football fans on Sunday afternoons in downtown Athens. In retrospect, without great risk, we can claim that these groups of security clerks and observers were the forerunners of what came to be known in the post-1981 period as PASOK's "green guards".

In the noise of this chaos, two of the three promised proposals, along with other selected interventions of delegates, were read. Papandreou's speech, the alleged event of the day, was little different from his speech to the Central Committee a couple months before. It was as usual, general, vague and skillfully delivered, so that at least on the surface the distance among the various currents in the organization was kept intact. However, we can identify three new elements in his logos which both anticipated the up-coming developments in the Movement and highlighted the shift in the organization's political priorities. More than ever before Papandreou tried to boost the membership's chauvinism. "We are the chosen", he argued, "for opening the new way...for the national liberation, the popular sovereignty and the social emancipation of the working Greek".[42] The implications of such a great mission were not left to the imagination of the audience. Given the difficulty of the project, "the numerous examples of lack of discipline and irresponsibility, especially by those who cry out in the name of democratic procedures...(will) not be tolerated. (To do) otherwise would be enormously irresponsible to the people who vested their trust in us".[43]

Secondly, Papandreou made, for the first time, a clear reference to the significance of his role in the organization and in a way he tried to present himself as the most accountable part of the organization. He did not dispute the argument that the "President of the Movement" should be elected by the membership in the Movement's Congress and that he should function according to the collective principles of the organization. But, he continued, "...his (the president's) role should not be limited to delivering speeches...since I have taken the initiative for the establishment of a Socialist Movement in Greece I have increased responsibilities to the membership and to our supporters".[44]

Finally, Papandreou made a deliberate effort to underscore the significance of the parliamentary wing of PASOK Critizing all those who wanted a socialist movement without a parliamentary wing – although to our knowledge, no political tendency in PASOK ever wanted anything like that – he argued that "to say a socialist movement without parliamentary expression is necessarily a revolutionary one (is) something which does not correspond to the thesis that the transition to socialism is possible through democratic means". In fact he went so far as to claim that parliamentary activity is as important as the activity outside parliament. "The parliamentary caucus has the responsibility for the immediate and short term issues... (while) the extra parliamentary bodies have the responsibilities for the recruitment and the development of the long term strategic plans".[45] It was the first time in the short history of PASOK, that not only was special reference and emphasis given to the importance of the parliamentary "struggle" but also, and this might be more significant, it was the first time that a separation between short term issues and long term goals was expressed.

But if Papandreou's proposal was controversial and created some scepticism on the part of the membership, the proposal on "The Organizational Principles of PASOK", by A. Tsochatzopoulos, who was later to become a prominent cabinet minister, was widely seen as scandalous and generated numerous reactions. It was a document which proposed not only a crude centralization of the Movement's structure but also the establishment of "the President" as a separate and all powerful "body" of the organization. The President would appoint the members of the Executive Bureau,[46] the Executive in turn would "create the Working Committees", the "Working Committees" would form the various bureaus, which in turn would nominate their chairpersons, who would finally be chosen by the Executive Bureau. In addition to this orgy of acclamations, nominations and appointments, which elevated the President to the only source of power, the proposal allowed some room for challenging the decisions of the Executive Bureau. This however would be possible only if a three out of five majority was achieved in the Central Committee!

The controversial nature of these two proposals and the reactions which they generated led the chairing committee of the pre-congress to adopt them as working documents for discussion as the Movement was marching towards its first congress. In fact the vision of the upcoming first congress, which incidentally was never denied even by the proposals themselves, somewhat cooled off the hot, but legalistically oriented heads of the opposition, and the Pre-Congress proceeded to the election of the Central Committee.

It soon became clear from Papandreou's statements (e.g. "we have to cleanse the Movement of the factions") as well as from the behind the scenes activities of the "President's men" (e.g. the circulation of the "preferred list") that the whole purpose of the pre-congress enterprise was the election of a Central Committee loyal to the President. But when the votes of the 500 delegates were counted, Andreas and his faithfuls realized that the plan had failed somewhat. The new Central Committee[47] was not freed from all those who had consistently opposed undemocratic practices and a role for the President beyond any collective control.

As we have mentioned repeatedly, PASOK was a meeting place for the radicalized activist. This radicalism, though more instinctual than conscious, was enough to resist the pressure of the leadership for complete submission to the Movement's structure. The results of the elections of the Central Committee highlighted this fact. However, the same results also illustrated that PASOK had started to take a rather decisive turn. All the members of the "opposition" who survived as members of the Central Committee had been elected with significantly fewer votes than the "President's men" or even than those members who chose to remain "neutral" in the dispute.[48] But this was not to be sufficient for the leadership. The events which unfolded with dramatic speed in the aftermath of the pre-congress became the living proof of this all or nothing attitude of the leadership. PASOK was heading towards its major split, which the reconciliatory results of the pre-congress could neither anticipate nor prevent.

The Split: The Beginning of the End

The failure of the leadership to put the organization under its complete control did not stop it from launching a direct attack aimed at the realization of its plans. Factors external to the organization (e.g. the press, the political practice of other political parties and institutions) were to play significant roles in this phase of the struggle over the Movement's structure, reminding us once again of the importance of the political discource/practice of the environment in the development of representative organizations.

The spark which set off the explosion that eventually blew the Movement apart, was once again set by Papandreou himself. On 21 March, a few days after the pre-congress, Papandreou announced to the press that the proposals were binding documents for the organization and not working drafts, as had been stated by the chairing committee of the Pre-congress. Putting aside for a moment the questionable nature of such an initiative, this would not have been a major problem, if the organizational proposal had not been included among the other proposals. The elevation of the organizational proposal to a binding document of PASOK implied acceptance of the "President" not as chairperson of the Central Committee – in any case Papandreou never subjected himself to the electoral "power" of the Central Committee – but rather as a "body" higher and parallel in power to the Central Committee.

This decision naturally generated reactions, especially among the membership which had all along argued against the centralization and personification of the Movement. But PASOK had not established the structural practices which would have allowed internal debates and the possible resolution of disputes. Thus the "opposition" and particularly members of the C.C. were left with the same options as rebels in a non-democratic regime: engaging in guerilla warfare and dreaming of the day of the big attack. So while they were waiting patiently for the first meeting of the C.C., at which the major attack was to be launched, they were leaking their reactions to the public through their personal connections with the press.[49] But even when Andreas finally decided to call the first meeting of the C.C. (April 13) they were faced with a surprise counter-attack which created a further set-back. Without any previous warning or discussion Papandreou announced "The Decision on the Internal Regulations of the C.C.", and declared them a "subject beyond the possibility of discussion". The move took the opposition by surprise and confused its original intent to attack the arbitrary adoption of the organizational proposal. Thus, instead of solving the existing dispute, they walked away from the meeting with an extra problem and a diffused focus since they now had to fight the "Regulations of the C.C." as well.

Despite appearances, PASOK's internal friction and disputes were not merely the product of differences in the politics of its top functionaries who voluntaristically decided to impose their will on the organization. Underneath all these were real issues concerning the actual day to day politics and development of the new Movement. Thus, it comes as no surprise that the unravelling of the "opposition" within the Movement began when the political/ideological differences clearly started to hinder PASOK's participation in mass politics. At the time, PASOK had an overwhelming presence among the youth and particularly in the students' movement. It was there that the internal dispute over the Movement's structure was translated into policy conflict, which in turn made the symbiosis of the two currents of the Movement impossible.

The first confrontation of the two currents took place around the issue of leadership of the National Students Union. The first post-dictatorship election of the National Council of the Union (May 1975) did not result in control of the executive by any of the existing organizations. Thus, *Panspoudastike* the student front organization of the C.P. and PASP (*Panellenea Agonistike Spoudastike Parataxe* – Panhellenic Militant Students' Front) the student front organization of PASOK put together an agreement and divided the positions and the control of the Executive. Many of the elected members of PASP, who had been identified with the opposition within the organization, claimed that the agreement, in bypassing the open democratic procedures of the Union, was creating a non-democratic precedent for the students' movement. These members, in spite of their diverse political and ideological attitudes, were strong believers in "democratic procedures and the omnipotency of the "mass movement". Consequently, they publicly denounced the agreement as an attempt at "wheeling and dealing behind the back of the students' movement". To them the "political autonomy of the students' movement should be preserved (if it) was to function with democratic procedures...something which in any case was part of the establishing declaration of PASP itself".[50]

PASP under the direction and the auspices of PASOK itself, reacted quickly to this "embarrassment". In an unscheduled meeting of the organization, fourteen members who had opposed the agreement were dismissed, five of whom were elected representatives in the National Union. It is interesting to note that *Panspoudastike* played a significant role in legitimizing this decision, reminding us once more of the difficulty of establishing democratic practices in an essentially hostile environment. Being a Communist Party front in the students' movement, *Panspoudastike* constituted by definition the most radical organized section in students politics. Thus, when the representatives of *Panspoudastike* applauded the decision from the forum of the Union's general assembly,[51] the decision of PASP to dismiss its rebellious activists gained legitimacy in the eyes of its uninformed and confused radical rank and file membership.

The dismissal of the opposition in PASP was followed by a series of internal debates and arguments centered on the question of the legitimacy of the dismissals. The students' organization of PASOK was deeply divided. Many sections of PASP, not rarely the majority in entire university faculties, openly disagreed with the decision and demanded its retraction and/or reconsideration. These debates often superseded the conventional notion of discussion and degenerated into fist fights between the pro and the anti-leadership groups. This growing dissatisfaction and disarray within the students' section of PASOK could have had unpredictable consequences for the Movement as a whole. Thus PASOK's leadership, which had already shown signs of intolerance to disagreements, intervened through the Youth Bureau, and dismissed all the critics. Every member of the Movement or of PASP who had disputed the dismissal of the fourteen members or even wanted a proper reconsideration of the controversial decision was told that by doing so they "were putting themselves outside the Movement".[52]

As one would have imagined, it did not take long for the turmoil in the students' organization to expand to the rest of the Movement. On 30 May, only five days after the first dismissals from PASP, twenty-one members of the Central Committee filed a protest to the Executive Bureau about the situation which was developing in the Movement. In this internal document they declared that it was unacceptable to be informed about the students' section of the organization through the press and called for a meeting of the Central Committee in order to discuss the problem.

Instead of an answer, the Executive appointed by Papandreou reacted with the announcement of the dismissal of eleven members of the Central Committee,[53] ostensibly because they were at the heart of the growing problem of divisiveness within the organization (4 June). At the same time, the Executive announced for 22 June, the meeting of the Central Committee at which the problem would be discussed thoroughly. The unfolding events however, made it clear that the upcoming meeting of the Central Committee would not simply ratify the decision of the dismissals. Amalia Fleming, a popular member of the resistance and well known as a supporter of Papandreou, and ten functionaries who had been members of PAK, stated their disagreement with the initiatives of the Executive. In an attempt to stop the rollercoaster of dissent, the Executive in a new memo to the organization, prohibited the discussion of the dismissals by the local and professional clubs and announced the dismissal of another thirty-seven functionaries out of which fourteen were members of the elected Central Committee.[54]

The attempt however failed momentarily, since a considerable number of local and professional clubs protested the Executive's decisions. Many clubs which had managed to overcome the prohibition of discussing the organizational turmoil went so far as to link the problem to the change in PASOK's overall policies. As becomes clear from the letters which these groups sent to the central bodies of the Movement, the dissident membership tried to link the organizational crisis with the growing tendency of the leadership to build the "public image of the Movement". In fact during this period two concurrent political events were put forward as examples of compromises of the alleged radicalism of the Movement. First there was Papandreou's flirtation with Willy Brandt, who had been visiting the country, and second the opening of the Movement's newspaper, *Exormise*, to corporate advertisement. More fundamental however, than the political objections of the "dissidents," were the accusations of undemocratic, often unethical practices of the Executive and Papandreou himself. Indeed many members found it unacceptable to accuse well-known "heroes" of the resistance, whose integrity no one had ever disputed, of being "divisive factionalists" objectively playing the game of the government, "elitists", "entrists" and the like.

Confronted with this situation, Papandreou, well aware of his exceptional position within the organization, decided to bypass the Executive and assume full responsibility for the organization. Obviously his prestige among the membership could not only legitimize the decisions of the Executive but could also minimize the organizational costs to PASOK. Thus he asked for and received, the resignation of the remaining members of the Central Committee (14 June), which in effect amounted to the official dismissal of the only elected collective body of the Movement. This was in fact the third Central Committee in PASOK's ten month life to be dismissed by its leader. The dismissal of the Central Committee marked the end of attempts to implement the "democratic procedures" principle of the Movement, and in effect the transfer of control of PASOK's structure to its President.

On 26 June, Papandreou issued an administrative order to all Committees of the organization which encapsulated this dramatic turn of the Movement and highlighted his success in taking over complete contol of the organization: "No order to the membership", he stated, "should be issued without my signature; (even) the participation in mass demonstrations, in defensive struggles, as well as in union elections etc. should be ratified by the respective committees, whose coordinators have to report to me".[55] The attempt to gain control was no longer only that – it was a *fait accompli*. That which had not been accomplished through the democratic facade of the pre-congress had now been achieved through administrative means.

Papandreou's initiative forced the opposition, which up to this point had been rather timid in its responses, (the main criticism of the "dissidents" was focused on the procedural question of the dismissals), to separate itself from the Movement's future. The organization thus experienced its first split. Hundreds of members signed petitions of resignation and numerous others simply left the Movement silently. This virtual avalanche of resignations was undoubtedly stronger in the urban centers where the most sophisticated and informed membership resided and less so in the countryside where the organization was very much under the influence of the *palaiokommatikoi*. In addition, a reading of the letters of resignation makes the diverse basis of the membership's disagreement more than obvious. The arguments of the members who left the Movement ranged from sophisticated predictions of PASOK's future to criticisms of the leadership on moral grounds, to even personalized disputes.[56] In this first major organizational turmoil, PASOK lost an estimated two thousand active members among whom were some prominent functionaries.

In spite of the magnitude of the resignations and the quality of the members who resigned, Papandreou declared that the split was an insignificant incident which was concentrated at the top. Though at the time such a statement sounded rather absurd, in retrospect the genius of Papandreou's political instincts were once more to be proven correct. The organization had virtually no official or even customary channels of communication. The only connection, to or between, the various parts of the Movement passed through the public domain i.e. the Parliament and the press. But these were channels that only Papandreou and/or the *palaiokommatikoi* had access to. The majority of the dissidents, including the dismissed members of the Central Committee had difficulties even uniting their voices and communicating. Even when a good number of them got together and formed a new organization – *Sosialistike Poria* (Socialist March)[57] – their only means of publicizing their views was through personal connections with some journalists and to a much lesser degree through the new group's small weekly newspaper.

This fact allowed Andreas not only to downplay the significance of the split but even to consolidate and expand his personal appeal and strength within and outside the organization. Skillfully using those two tendencies which were loyal to him (i.e. PAK and the *palaiokommatikoi*), he managed to communicate his own interpretation of the split and the politics of the dissidents. Thus in addition to his insistence that the crisis of the organization was confined to the top, he managed to propagate the thesis that the dissidents were either social-democrats with elitist beliefs or communists who wanted to appropriate the Movement for their own purposes. On the one hand, the conception of the dissidents as elitist social democrats spread widely among the rank and file of the urban and professional branches of the organization, who were allegedly fairly radical and quite critical of the leadership's initiatives. On the other hand, the conception of the dissidents as opportunistic communists became the conviction in the more conservative and uninformed countryside through the deployment of the clientilistic network of the *palaiokommatikoi*. In this way Papandreou managed not only to legitimize his initiatives and isolate the dissidents, but also to expand his public appeal, particularly in the countryside.

Thus as we will see, after a short period of confusion and organizational stagnation, and contrary to the predictions of the writer and the majority of PASOK's ousted members at the time, it became evident that Papandreou had brilliantly turned the first major crisis of the organization from a reason for contempt, to the basis for a leap forward.

The Response

By underlining Papandreou's key role in the events which preceded and followed the split, we did not intend to reduce this tragic development of the Movement to the mere initiative of one man. Such developments in political organizations are the complex outcome of a number of interrelated factors and relationships in and outside their structural/organizational confinements. The role of the individual(s), even if we are dealing, as in our case, with an exceptionally powerful one, can only be fully understood in the context of these other factors. Thus, Papandreou's brilliant political manoeuvring would not have been so successful if it had not been for the particularities of the various tendencies within PASOK, and the conducively silent role of the other political parties (especially those on the left), in addition to the changing political climate.

The reactions of the various tendencies of the Movement to the dramatic developments within the organization varied according to their strategic vision or short term politics. Thus, the hard core of PAK, which, as we have said, maintained a somewhat separate existence within the organization as "PASOK B", strongly supported and in fact promoted the leadership's initiatives. Having a rather strong Leninist conviction, this current firmly believed that by ousting the membership which argued for the institutionalization of democratic procedures (the backbone of which was D.D.), that they (PAK) would be able to take over the Movement and organize it along vanguard lines.[58]

On the other hand, the *palaiokommatikoi*, composing the overwhelming majority of the parliamentary wing of the organization saw the dismissals as an opportunity to improve their rather inferior status in the Movement. The withdrawal of those members with the strongest feelings for collective democratic procedures and against traditional organizational patterns (clientelism) could only mean a better future for the *palaiokommatikoi*. Thus although they did not vocally support the reasoning of the leadership – in any case that would have been difficult given Papandreou's close ties with the radical ex-PAK membership – they quietly accepted the decisions of the leadership. Their occasional objections to the purges were not only limited to, but in fact focused primarily on, the procedure followed, and were never elevated to principled opposition.[59]

Finally, the Trotskyists remained suspiciously silent when confronted with the orgies of arbitrary decisions made by the leadership. Although some criticized the initiatives which led to the purges, their criticism, in a fashion similar to that of the *palaiokommatikoi*, was not made on theoretical grounds concerning the implication of the procedure and practices of the leadership. Rather it was made on the details of those procedures and practices which they often personalized to a very high degree. Furthermore they confined their "opposition" to some discussions at local clubs. However, as there was not a proper structural connection between the rank and file and the leadership, or among the local clubs themselves, even this mild criticism was condemned to be marginal. Thus it is rather safe to claim that the Trotskyist tendency remained sadly "neutral" during the whole period of the controversy surrounding the split.

The explanation for their reaction must lie in the fact that their presence in the organization was completely subject to Papandreou's whims. Many of them occupied high positions in the Movement's structure and being committed to an entrist strategy they did not want to jeopardize the possibility of its success. In fact some of them appeared to be making significant advances since a number of their well known members were appointed by Papandreou himself to key positions (e.g. coordinators of the "Organizing Committee" and of the "Center of Studies and Enlightenment – KE.ME.DIA.) and their newspaper – *Xekinema* (the Beginning), which had just come out (May Day 1975) seemed to be tolerated by PASOK. It was these considerations which led this otherwise radical tendency within the organization, not only to downplay its objections to the Movement's dramatic developments, but also to openly accommodate the leadership, as in the case of the voluntary resignation of the three Trotskyist members of the Central Committee, which legitimized Papandreou's dismissal of the organization's only elected body.

While all these short term calculations of the Movement's various groupings and currents legitimized the leadership's actions within the organizaton, the response of the other political parties was critical in doing the same from outside. Political parties are determined as much by societal/political patterns, and the nature of the practices/discourse of their competitors i.e. other political parties, as they are by internal factors. The Greek party system, as we have seen in chapter I, was characterized by anything but democratic practices. The parties' memberships really existed only to carry out the decisions taken by the leaders, as in the case of right wing parties, or by the politbureau, as in the case of the left. This alone was enough to explain the virtual silence of the other parties (evident in their newspapers and journals) to PASOK's internal conflict. When undemocratic practices occur in a non-democratic environment, it is only natural to assume that such practices will not be opposed but will in fact be legitimized.

In addition to the fact that the practices developing in PASOK were not uncommon in the context of Greek politics, the actual reactions of the left parties further legitimized Papandreou's initiatives. We have already seen the encouragement which the students' organization of the C.P. gave to PASOK during the purges in the students' movement. The C.P.s declared that the purges were exclusively the internal affair of the Movement and remained provocatively silent. Of course besides the obvious explanation that a democratically run radical socialist Movement would have jeopardized their domain, there was also the fact that from the beginning the C.P.s had trouble coping with presence of another left organization in the political spectrum. As Poulantzas later put it, the left had failed to confront and criticize PASOK "as a left socialist movement and not as...a mythic populist (one) despite the ideological implications of the term which conceals the suspicion of 'bonapartism' (we might have)...regarding its leader". "For anyone to deny", Poulantzas continued, "(PASOK's) left socialist character to the extent that it materializes within it – as the other left parties in our country – the peculiar coexistence between factions of the working class, the peasantry and the radicalized petty bourgeois strata is a huge mistake and implies a completely mechanistic conception, as far as the relation between social classes and political parties (is concerned)".[60] Therefore this understanding, or rather the lack thereof, regarding PASOK, resulted in a rather superficial and in effect arrogant attitude on the part of the C.P.s *vis-a-vis* its internal developments, and further facilitated and legitimized its unfortunate outcome.

Finally, the changing atmosphere of the political environment was conducive to PASOK's leadership's initiatives. The right's overwhelming victory in the 1974 election resulted in a rough landing for some leftists within the Movement who had overestimated the post-dictatorship radicalism. In turn, as time passed and it became clear that socialism was not imminently on the horizon, the desire for political effectiveness and perhaps efficiency among the inexperienced and anxious membership became widespread. Also contributing to this desire was the cooling off of the post-*metapolitefse* radicalism which became more and more evident as the right, bypassing the first turbulance of the political change, started to reorganize the state without real opposition. This realization made the question of democracy within PASOK seem a luxury item, a counter demand, and made its advocates appear, at the very least, suspicious.

Conclusion

In conclusion, we can safely claim that the immediate post-1974 election period was for PASOK the most significant in its development. During this period, the organization developed a set of new practices which were to become the permanent *modus operandi* of the Movement, and which in turn set the basis for its subsequent political transformation. PASOK had begun as a promising radical socialist movement with the potential to become an organized socialist party which could live, develop and eventually capture power using its leader's charisma, but functioning independently from his will. The split, and especially the events which preceeded it, indicated that this potential was to be cut short. The conflict between those in the organization who argued for a democratically run PASOK, and the leadership (A. Papandreou and those members loyal to him) who wanted to put the Movement under its exclusive control was resolved with the victory of the latter.

The 1975 split, contrary to other contentions, indicated that PASOK was the creation of A. Papandreou and that he was ultimately its sole source of power. From the beginning the Panhellenic Socialist Movement was not "up for grabs" by the various political currents it had attracted. It was rather an organization whose political power was determined by Papandreou and the political conjuncture at the time. This bitter truth was something that the "non-Andreist" membership failed to understand. Of course this is not to say that from the beginning there was no hope for the success of them. Rather it means that a better understanding of the actual political game being played out in the organization would have helped these members to "play their cards" to better effect. Instead of assuming that it was sufficient for them to try to confine their efforts at influencing the Movement internally, they should have realized that their road to success had to pass through participation in the political game (i.e. elections, old political practices). Their absence from actual public life both facilited Papandreou's attempts at complete control of the organization and further isolated them during the organizational crisis.

The opposition was correct to focus its efforts on the organizational question. But it did so without realizing that its existence as distinct political currents depended on the will and the tolerance of the power source of the Movement, namely Papandreou himself. The continued existence of PASOK's internal tendencies depended on Papandreou as much as the fetus depends on the mother for its survival and development. The only way to independent existence lay within the confines defined by Papandreou and the environment. Instead, those tendencies in opposition to Papandreou appeared to have cut the umbilical cord since they tried to gain strength according to ideal conditions (which did not exist) and not through the existing rules of the political game of the time (e.g. electoral competition) and the Party's internal balance of power.

In addition, currents opposing the leadership were anything but unified, since the young Movement had not even developed a language of its own. This not only reduced the opposition's effectiveness but also came in handy to Papandreou and his friends who skillfully managed to play off one political current within the Movement against the other. In fact, this Disraelian tactic was to become, as we will see, a favourite practice of the leadership as it further attempted to define the Movement's future in its own terms. Soon even the tendencies which had aligned themselves with the leadership in the 1975 organizational crisis were to find themselves under the pressure of the same tactics.

To sum up, the 1975 crisis and split in PASOK represents a key turning point in its development. The developments in PASOK during this period are an excellent example to illustrate the importance of the connection between the organizational structure of a political party and its politics. Although the factional war within the Movement was not fought along explicitly ideological and political lines, there is no doubt that it paved the way for the organization's political transformation. The potentially radical socialist Movement ("of protest") was indeed to become the legitimate leading party of the opposition and Papandreou would soon exchange his turtle neck sweaters for silk ties.

Notes

1. From the *Hypoepitrope Organotikou Grafio Eparchias* (Report of the "Subcommittee of Organizing" – The Bureau for the Provinces), December 16, 1974. Source: Personal Archives of Members of the Central Committee of the Party": Andreas Politis; Gerasimos Notaras.

2. Ibid.

3. Proposal by G. Notaras on the *Katastiko tes Kentrikes Epitropes* (Constitution of the Central Committee), *Mimeo*. Source: Personal Archives.

4. *Anafora kai Eisegeses* (Report and Proposals). Report of the Trade Union Bureau. See also pamphlet edited by the same bureau: *PASOK: For the Workers*. Source: Personal Archives.

5. Interview with G. Tzafoulias, Secretary of PASOK, who lived in

Florence Italy until 1977, July 1984. The same conclusion is also derived from the various political resolutions of the Italian organizations of the Movement: 29-30 November 1975 and 11 December 1976, *Mimeo*. Source: Personal Archives.

6. Ibid.

7. Many of the alleged members of this faction not only were members of the Executive Bureau appointed by Papandreou but they were considered part of his inner circle [Tsohatzopoulos, Tsekouras, Vaso Papandreou (no relation to Andreas), Semites].

8. See for example the first issues of *Agonistes*, the publication of the youth section of PASOK, as well as the pamphlets published during this period: *PASOK kai Koinonikes Taxes* (PASOK and Social Classes), *Yia to Socialismo* (On Socialism), *Ti Thelei to PASOK?* (What does PASOK Want?).

9. See: *Syndagma yia mia Democratike Ellada* (Constitution for a Democratic Greece) and *Oi Theses Mas: Democratiko Panepistemio, Pragmatike Ekpaedefse, Yia Mia Nea Zoe* (Our Positions: Democratic University, Real Education, For a New Life).

10. For example, he never even made the symbolic gesture to participate in any of the sub-committees or Bureaus of the organization. His presence in some of them was much desired given his academic background and international connections.

11. In fact Papandreou never became a member of the Central Committee of PASOK. It was always the "Central Committee and the President" or the "Executive Bureau and the President". Even today when he is in complete control of the organization he maintains the same pattern.

12. Internal Document-Resolution of the Provisional Central Committee "Yia ten Anaptexe tes Organoses tou Kinematos" (For the Advancement of the Organization of the Movement). December 1974 (precise date unavailable) p. 1. Source: Personal Archives of Certain Members of the Party's Central Committee: A. Polites and G. Notaras.

13. Interviews with G. Notaras and A. Stangos op. cit.

14. A. Papandreou, Speech to the Central Committee, 19 January 1975, in Andreas Papandreou, *Apo PAK sto PASOK* (From PAK to PASOK) Athens: Ladias Publications, 1976. p. 86.

15. Ibid. It is worth noting that this kind of argument, which in effect waves the label of treason over the heads of the party-line critics, is not unknown in the history of the Greek left.

16. Ibid. pp. 86-87.

17. Ibid.

18. Ibid. p. 90.

19. Of course it is rather redundant to say that the left never advocated undemocratic procedures for gaining power.

20. Ibid.
21. Ibid.
22. Ibid.
23. This was further assisted by the fact that ideological and political debates had not yet permeated the rank and file membership, especially in the countryside.
24. This expression was introduced by Papandreou and the Executive Bureau and was to become the standard phrase used during numerous purges from the organization until today.
25. Source: Interview with Stangos, op. cit.
26. Interview with G. Notaras op. cit.
27. This is what Papandreou claimed when he was confronted by critics of his decision.
28. From the President's announcement on the "Cancellation of the functions of the Provisional Central Committee". *Agonistes* 7 February 1975.
29. The members Karagiorgas, Manolkides, Stangos, Filias and Semites (who later became Minister of Agriculture and then Minister of National Economy) opposed the decision, while Mandelara, Tsouras, Tsohatzopoulos and V. Papandreou supported it.
30. *Agonistes* February 7, 1975.
31. L. Philippatos "Gramma ston Proedro" (Letter to the President) March 5, 1975. *Mimeo* p. 5-6. Source: Personal Archives.
32. Short papers with themes such as "Democratic Procedures and Socialism", "The Relations of the Political Party to the Mass Movements", "The Importance of Horizontal Connections between the Clubs of a Socialist Organization" were written and, debated. As in the proposals of the Engineering School, the Law School, the Physicists and the Ilisia clubs, *Mimeos* Source: Personal Archives.
33. Interestingly enough, five out of the fifteen members of the Committee were among the most outspoken opponents of the decision. Source: Personal Archives.
34. This proposal was never delivered. Personal Witness.
35. Interviews with: Stangos, op. cit., Notaras, op. cit. The same point was made by many activists within the party during the split. Personal Witness.
36. It is worth noting that although in these counter proposals/ protests A. Papandreou was implicated and critized, he was never mentioned by name.
37. From the "Paratereses tes Topikes Organoses Illision" (Remarks of the Local Club of Illisia). 14 March 1975, *Mimeo*. Source" Personal Archives of Certain Member of the Central Committee of the Party: A. Polites; G. Notaras.
38. Karagiorgas, Manolkides, Stangos, Filias.
39. Memo signed by "The Executive Bureau". 14 March 1975.

Source: Personal Archives of Certain Members of the Central Committee of the Party: A. Polites; G. Notaras.

40. The participation of the opposition members in the "Preparatory Committee" was rather nominal, which allowed a lot of room for the cultivation of polarization and fanaticism within the Movement. Thus, these rank and file members were usually chosen by local clubs, which had loose ties with the headquarters and therefore were more vulnerable to misinformation. For example, the agreement between D.D. and PAK, which was obviously put in jeopardy remained unknown to the rank and file for some nine years after it was first published. See chapter II.

41. Two episodes characteristic of this kind of activity infused some scepticism among the uncommitted members about the leadership's initiatives. The first was the violent physical confrontation of these groups with the resigned members of the *Agonistes* editorial board who tried to distribute four short commentaries on the Pre-Congress; and the second when the late D. Karagiorgas, a prominent member of D.D., member of the Executive Bureau and a well respected "resistance figure" – he had been crippled during the resistance – tried to intervene during the procedure, he was shouted down and crassly called names such as cripple and worse.

42. A. Papandreou "Proposal – Speech at the Pre-congress". March 16, 1975. *Mimeo* p. 1, also in A. Papandreou, *Apo PAK Sto PASOK* (From PAK to PASOK) Athens: Ladias, 1976.

43. Ibid. pp. 1 and 6.

44. Ibid.

45. Ibid.

46. The President was not even bound as to the number of the members of the Executive Bureau. The proposal suggested a number between nine and thirteen.

47. See Appendix II.

48. Results of the elections of the Central Committee *Agonistes* April 9, 1975.

49. That was particularly the case with the C.C. members A. Stangos and P. Eftimiou. Interviews with G. Notaras and A. Stangos, op. cit.

These initiatives were used as an excuse by the leadership to ostracize the opposition later on. In addition, they contributed to the alienation of the opposition from the rank and file the latter of which considered these activities uncomradely and highly un-disciplined.

50. From the "Proposal of the five dismissed members of PASP" to the National Students Union. *Mimeo* p. 1, Source: Personal Archives.

51. In fact "Panspoudastike" went so far as to claim that PASP should have dismissed these members earlier. Personal witness.

52. This phrase was to become a trademark as the leadership was to

The Critical Year of 1975

make extensive use of it when the organization was to undergo a virtually continuous series of splits.

53. There were 6 members of the old D.D., 2 members of PAK and 3 members of independent left origin. Source: Personal Archives of Certain Members of the Central Committee of the Party: A. Polites; G. Notaras.

54. Memo No. 23, 28 June 1975. This brought the number of the dismissed members of the C.C. up to 35 out of the total membership of 75. Source: Personal Archives of Certain Members of the Central Committee: A. Polites; G. Notaras.

55. Memo/order: From the President to Members of the Committees. 26 June 1975. *Mimeo.*

56. This is evident not only from the diversity of the focus of the numerous letters of resignation but also from the diverse political orientation of the resigning membership. In retrospect this fact partly explains the problems which the opposition had in forming a coherent political expression.

57. On 7 July 1975 approximately one thousand ex-members of PASOK gathered in a downtown theater in Athens ("Satira") and established the "Socialist March". The organization called itself a "movement" (kinesis) and adopted as its primary strategic goal the "unity of the renaissance and the unity of the left", "A´ Conference 13-14 December 1975". [See: *Apophase tes A´ Pan-hellenias Syndiaskepses* (Resolutions of the First Panhellenic Conference) Athens: Sosialistike Poria, 1976.] The organization had a rather hectic development and survived until 1979. Although it managed to attract a good number of independent leftists, it did not become a mass organization – never managing to attract more than two thousand members – nor did it become a counter-pole of attraction to Papandreou's PASOK, as was at least initially implied. Thus, "Socialist March" became the symbol of the impasse which the ideas of the new left met in the conjuncture of Greek politics.

58. In February, 1975 a few months before the actual split, the branch of PASOK in Switzerland, dominated by PAK members published an article in *Agonistes* in which the purging of the "social-democratic faction of Democratic Defence" was argued. Similar decisions appeared to be reached by the conference of the Italian section of the Movement during the same period. From the Resolution of the Conference of PASOK of Italy. *Mimeo.* Source: Personal Archives of Certain Members of the Central Committee of the Party: A. Polites; G. Notaras.

59. The mild and conjunctural nature of the *palaiokommatikoi* opposition explains their fast reconciliation with the leadership soon after the turmoil was over. The cases of M. Merkouri and A. Fleming can be cited as prime examples of this syndrome.

60. N. Poulantzas, "What can be done with the unity of the Powers of Change" in *TA NEA* February 1978 and also in P. Sarantopoulos (ed.) *PASOK kai Exousia* (PASOK and Power), Athens: Paratirites, 1980.

4

From Movement of Protest to Leading Opposition Party 1975-1977

Despite the conflicts within the Movement in 1975, PASOK proceeded apace in its short march to power. However, by 1977, the Party had dramatically changed its image. This shift was not simply the result of the deliberate manipulation of the vagueness of the Declaration of the 3rd of September by the leadership following the 1975 ostracization of the opposition. This moderation of the Movement had come about as a response to the post-*metapolitefse* political and economic changes in the country as well as the series of internal crises over the Movement's organizational structure which had further strengthened its leadership. In the 1977 election, PASOK presented itself to the electorate no longer as a "party of protest" as in the 1974 election but rather as a serious party capable at least, of conducting the opposition. And as the Movement's electoral lists were flooded by liberal opportunists fleeing the sinking parties of the centre, even party sympathizers had a hard time distinguishing in any clear fashion PASOK's practices from those of the *palaiokommatikoi*, which the Movement had previously denounced. The Movement had changed dramatically, launching itself on an apparently irreversable course. But this had not happened overnight. It was a long way from "September 3rd, 1974" to the election of 1977. In the following pages we will examine the external factors and the internal developments which were conducive to PASOK's transformation.

A Conducive Environment – The Government of New Democracy

The post-1974 government of Konstantine Karamanlis and his newly established party bore little resemblance to either the conservatism of the pre-junta right or the neo-liberal leanings of its European counterparts. From 1974 to 1977, New Democracy, under the strong leadership of Karamanlis, managed to undermine some of the ultra-right elements in its ranks and present itself as a modern, open, liberal party committed to "the principles of "social democracy, political freedom and social justice which are identical with one another".[1] Even further, the Government Party held that "political freedom can only be guarenteed if there is a fair sharing of national income".[2] Although these commitments can be seen as rhetorical and tactical manoeuvres in the context of wide-spread radicalism in the post-*metapolitefse* Greece, there is little doubt that the Karamanlis government set its face against orthodox economic policy and did not hesitate to carry out nationalizations and expand state investment in the secondary sector.[3]

With respect to the country's foreign policy, the government of New Democracy displayed an admirable degree of liberalism by Greek standards. Although operating under its leader's dogma "We belong to the West", the Karamanlis government did not hesitate to respond to the wide-spread questioning of Greece – US relations by the spectacular move of withdrawing the country's membership from the military wing of NATO (although it did not of course follow through consistently with the results of this initiative) and in addition followed a "multi-dimensional" foreign policy. The latter was expressed though the continuation of good relations with the Arab world,[4] as well as in numerous visits to Eastern Europe, the Soviet Union and even the People's Republic of China. It was an initiative that the left opposition did not fail to support, since it was undermining the traditional anti-communism of the country's political discourse.

Always with its commitment to the country's membership in the EEC and its overall strategy for the consolidation of a liberal democratic system as a background, the government's liberal intentions were expressed in a number of intiatives covering a wide spectrum of issues. These included the legalization of the C.P.s, the relaxation of the internal surveillance that the Government had inherited from the days immediately prior to and during the Junta, language reform, the positive response to some labour demands such as the reduction of the number of working hours per week,[5] at least the symbolic consideration of women's issues,[6] and some modernizing measures regarding the archaic educational system. On the economic front during these first years of the *metapolitefse*, the rates of economic growth were rather impressive, especially when we consider the relatively slow start in 1974 with 3.4%. The rates of economic growth were 6% in 1975, 6.4% in 1976, and 3.4% in 1977.[7] These growth rates were largely attributable to the growth of industrial production, for which the average rate during the period was 5.95%.[8] Overall, the economic policies of the first Karamanlis government were of a modernizing nature, and labour, under the auspices of governmental corporatist intervention in labour relations, and due to its own radicalism, managed to make considerable gains in real wages and a reduction in working hours.[9]

Furthermore, during the 1974-1977 period we witnessed the crystallization of the demographic and social effects of this economic growth. Although the mass exodus from the countryside had already taken place in the 1960s, the trend continued. More than 1.5 million people (out of a total population of 9 million) moved to the urban centers during the 1970s.[10] When we take into account the anarchistic way in which these demographic changes took place, it is not difficult to imagine the numerous problems and frustrations which the city dwellers had to face. The chaos which is the city of Athens is a case in point.

In addition, a comparison of employment at the beginning and the end of the 1970s[11] clearly indicates the establishment of a new pattern which can be summarized in the following trends. First, although the traditional urban petty-bourgeoisie (statistically the self-employed and their unpaid family members) continued to be numerically sizeable, they experienced some decline, making up approximately 15% of the active population. Secondly, the wage and salary earners who composed the majority of the new urban working class strata were on the rise, and for the first time were rapidly approaching 50% of the active population (i.e. approximately 46%). In spite of the gains in real wages, the consolidation of this social trend, in combination with the continuous significant decline in the rural population resulted in the growing political frustration of these strata. This new situation signalled in effect the beginning of the end of the traditional petty-bourgeois dream of becoming one's own boss.

The political frustration of the new urban working class and of the peasantry started to become clear during this period through the nature and the intensity of the often spontaneous mobilizations which took place.[12] These mobilizations arose in protest against the Government's attempts to further "modernize" and "rationalize" the economy, and to increase productivity in order to respond to the capital needs engendered by the imminent entrance of the country into the EEC. Thus by 1976, the Karamanlis government begin to cool off its liberal policies. It attempted to control and even to reverse the opposition's attempts at democratization of the chronically ill state apparatus as it was expressed by the process of *apojuntopoiese* (de-juntaization) – i.e. the removal of the pro-junta elements from the state apparatus. In addition, the Government introduced a number of restrictive pieces of legislation *vis-a-vis* the mass movement. The infamous anti-Labour Act 330/76 became the landmark of these efforts.

PASOK responded to these issues with consistent programmatic promises of the democratization of public life, protective labour legislation, and reform of the dismal situations in health care and education. It also expressed concern about the isssues arising out of rapid urbanization such as pollution, traffic, housing, etc. These positions championed by PASOK soon became the most visible points of differentiation between it and the Karamanlis government and the basis of its growing appeal among the social strata protesting the Government.

The political, social and economic developments of the first post-*metapolitefse* years were conducive to PASOK's movement away from its original radicalism in several ways. On the one hand, the presence in Government of a moderate right-wing party (centre-right but more centre than right) was forcing PASOK to respond programmatically not with the vague radicalism of 1974, but rather in a concrete, "serious" and technocratic fashion which would address the solution to the country's problems. Of course this does not mean that PASOK or any socialist party for that matter could not both respond programmatically to the country's day to day problems and maintain a strategic socialist vision, but in the Movement's case, its internal organizational developments, as we will see below, circumvented that possibility.

On the other hand, the nature of the new wave of political

mobilizations was generally one which was permeated by a short-sighted petty-bourgeois flavour. Due to the absence of a strong working class tradition, it was led by the new urban working class and populated by a radical social strata of diverse origins including a strong petty-bourgeoisie element. This growing petty-bourgeois attitude was expressed in the "classless" and the "anti-government" base of this radicalism, the use of traditional forms of protest (e.g. it is not by accident that most protest ended up in front of the parliament or a minister's office where a delegation presented the protesters' demands), the individualist and sectoral nature of demands and of course the lack of any collective vision except the demand for the immediate change of government.

Finally, this environment was not only conducive to PASOK's political moderation but also to the undemocratic, centralized and electorally oriented development of its organization. When gradually but steadily the popular opposition to New Democracy's government turned its support to PASOK, given the virtual absence of experience in democratic party organizations or in democratic collective mass participation patterns, it remained indifferent to the Movement's leadership's manipulation of PASOK's structure or to the concerns of the Movement's radical/socialist tendencies, which were to be forced out of the Movement. But before we turn our attention to the internal events of the organization, it is worth looking at the overall attempts of the leadership – in addition to the 1977 electoral programme – to create a moderate image for the Movement.

Changing Politics – Changing Image

Until the spring of 1975, PASOK had the reputation of a radical party of protest without a legitimate, "serious" plan to govern the country. It was seen as the "big mouth" party on the Greek political scene, which had become the foremost expression of the post-*metapolitefse* radicalism. With such an image, PASOK could not be considered a governmental party with serious chances of capturing power. This became more clear as the post-*metapolitefse* radicalism started to display its lack of depth and began to simmer down.[13]

From Movement of Protest to Leading Opposition Party

After the spring of 1975 however, we started to witness a change in the Movement's political attitude. And just as up to point PASOK's public image was primarily created by its leader, it was Papandreou again who initiated the organization's new political make-up. Papandreou began to make himself available to the media much more than before and this time without discriminating against the politics of the newspapers. In the numerous interviews he gave during the period under examination, Papandreou appeared as a calm, "rational" political leader – without the radical jargon of the 1975 electoral campaign, downplaying his (up to that point) familiar "antis" – and focusing more (and at times exclusively) on the possible solution of day to day problems.[14]

In addition to this attempt at the creation of a new image for the Movement, PASOK started to take initiatives which went far beyond symbolism, generating an atmosphere of toleration if not acceptance, not only by the "general public" (electorate), but by the functionaries of capital. Thus in the summer of 1975 the parliamentary wing of the Movement voted along with all the other parties – the C.P. was the only exception – in favour of the updating and expanding of the privileges of foreign capital (Act 4171/61).

This does not mean however that PASOK had suddenly completely abandoned its initial politics. Rather it means that the organizational dismorphy of the Movement was creating a PASOK of two speeds, of two distinct (and not rarely controversial) practices and rhetoric. The phenomenon was shaped after the 1975 crisis and gradually grew to become the trademark of PASOK's political practice. On the one hand the first speed was composed of a radical (not rarely Marxist) political discourse which remained internal to the Movement and it found expression only in the party's internal documents and press. Of course some of these writings were public as in the case of *Exormise* but judging from the distribution numbers as well as from the fact that PASOK had not formally and explicitly recognized the newspaper as its own, the audience of this radical discourse was at best made up of committed activists of the Movement, who in any case tend to be more radical.

On the other side of this practice there was a significantly more moderate political discourse – the second speed, which was sustained through the public initiatives of the Movement in its parliamentary wing and especially by Papandreou himself.

By virtue of its public nature, this second speed was by definition superseding the discourse which was confined within the framework of the party structure. Thus for example during Giscard d'Estang's visit related to the country's membership to the EEC, while *Exormise* (PASOK's newspaper) headlines read "The Government undermines National Sovereignty: Sell-out to the EEC",[15] Papandreou's comment during the reception of the French president was: "Of course we are the opposition but we support Karamanlis' politics *vis-a-vis* France and the EEC". It should be pointed that Papandreou was part of both these contradictory practices of the Movement but undoubtedly, as the pillar of PASOK's public image, he contributed more to the latter than to the former, although of course he was not absent from the more theoretical and radical debates internal to the organization.

The juxtaposition of the politics promoted by the party functionaries against the politics deriving from some incidents involving the Movement's parliamentarians make this practice of the "two speeds" seem even more vivid. While for example the Movement's publications were devoting long and often special issues to the glorious history of the World War II Greek resistance movement and/or to attacks on social democracy, the *palaiokommatikoi* made what could be termed, at the least, controversial statements on the same issues. At the same time, I Alevras, a prominent member of the *palaiokommatikoi* and the "parliamentary representative" of the Movement used traditional anti-communist language during his testimony at the Junta trial[16] and other M.P.s were putting forward the "Swedish model of socialism" as an inspiration for PASOK.[17]

This two speed tactic contributed to the further projecting of Papandreou's personality as a legitimate alternative to the undisputed personalized power of Karamanlis, and in the process, to the changing of PASOK's image from a movement of protest to a viable alternative to the government. The former should in fact be seen as very important in the context of Greek politics in which the political parties did not have a long history and were usually identified with their leaders. This much more mainstream legitimization of Papandreou as a personality can be demonstrated in the importance Papandreou started to pay to his regular special sessions with Karamanlis as he sought to solidify his association with this popular political personality. In fact Papandreou began to develop such good rapport with

Karamanlis that their relationship started to take scandalous dimensions when at times he seemed to virtually ignore the Movement's existence. For example, after attending the week long Conference of Socialist Parties of the Mediterranean in Malta (18-25 June 1976), and before even meeting with his party's functionaries, Papandreou rushed from the airport to Karamanlis' office to inform him about the Conference.

This changing of PASOK's image did not necessarily imply a radical change in the Movement's politics. The Movement maintained a strong anti-monopoloy, anti-imperialist discourse. In fact the anti-imperialism of the Movement as it was exemplified by Papandreou's controversial stands, often went too far. Thus, Amin's regime was praised as a "great anti-imperialist force in Africa", the Ba'ath Party of Syria was to become a fraternal organization with PASOK, and Kadafi's regime was seen not only as a genuine anti-imperialist force in the Mediterranean but also as "the most democratic" and in "pursuit...of the classless society".[18] Strong stands continued to be taken against the country's membership in NATO and the EEC and on the need for structural change of the Greek society. However, the emphasis upon various elements of the organization's politics were different now. For example we can see an even stronger emphasis on the nationalist nature of the Movement's politics as well as a further concentration on the parliamentary way of conducting politics. Papandreou's call for "Hellenization of the state apparatus" as the basis for the proposed "unity of the democratic forces" (against the right),[19] the chauvinistic pledge to respond violently Turkish provocations in the Aegean,[20] and the repeatedly opportunistic and/or naive gestures with regard to the armed forces and their role in the country[21] along with the emphasis on the importance of "battles in parliament"[22] are some examples of this shift.

However, all these were not enough to complete the new image that the leadership of PASOK appeared to want to develop. Significant elements within the Movement continued to believe and act according to the original premises of PASOK. (These premises had never of course been openly denied). They still saw PASOK as an attempt to build a mass, democratically run, genuinely radical, even Marxist socialist Movement, and they were able to distort the developing new image of the organization. Therefore, in addition to its public initiative, it became necessary for the leadership to take action and cleanse the

Movement's ranks of these potential trouble-makers. Thus both the Trotskyist tendency and the militant part of the former PAK had to go.

The Organization

Following the major split which had occurred in its ranks in the spring of 1975, and as the leadership struggled to revamp the organization's image, PASOK was to experience another two years of internal turbulence. In the immediate aftermath of that spring, PASOK experienced further resignations, a general halt in recruitment, and overall disarray within the organization. The leadership responded by announcing a series of organizational activities and ideological debates. It soon became obvious that certain disruptive tendencies within the Movement who still viewed PASOK as an attempt to build a mass, democratically run radical socialist movement would soon interfere with the development of the Movement's new image. Consequently it was deemed necessary to expel both the Trotskyist tendency and the militant part of the former PAK. It was at the end of this second crisis of expulsions that PASOK entered its first pan-Hellenic Conference finally free of internal opposition.

Regrouping after the Split

A great feeling of uncertainty grew in the organization as the storm of resignations continued even until the end of September 1975. In addition, the members of the newly formed *Sosialistike Poria* (Socialist March – S.P.) were criss-crossing the country and using their personal contacts to lobby the membership of the Movement to join them. There was an almost complete absence of recruitment to the Movement, the "confusion and disappointment were beyond any doubt",[23] while often sections of the organization proceeded with initiatives which were strongly opposed by the central "authorities" of the Movement. It was particularly the latter which gave the impression of PASOK as an organization in disarray and made its future appear questionable.[24]

The reponse of the leadership to this dismal state of affairs in the Movement was the announcement of a series of organizational activities "for a stronger PASOK" and the promotion of intense ideological "debates". It was no secret – although never openly admitted – that the recently dismissed opposition as well as the new developing discontent were coming from the "left" or more specifically from those who believed in the development of PASOK along the lines of a mass socialist (though never clearly defined), democratically run Movement. The promotion of both intense organizational activity and ideological debates appeared to be the obvious answer to these concerns. In fact this strategy of intense ideological debates and organizational activities in the aftermath of crises, was to develop, as we will see, into a standard practice every time the leadership's choices were confronted with new challenges.

After the June crisis, A. Papandreou announced that a new Central Committee of the Movement was to be elected through a "new appeal to the rank and file" and that it was to have its first general meeting by 13 July. A few days later however, he announced that from the "proposals of the rank and file" only the "best" were to be appointed to the new central bodies of the organization, of which there were to be two: the Central Committee and the National Council. Again however it did not take long for Papandreou to change his mind. In another new memo,[25] he announced his plan for "Regional Conferences and the (first) National Conference" of the Movement. The organizing of these Regional Conferences presupposed a number of unclear and chaotic procedures which were to be implemented in a short period of time and under the strict supervision of the leadership's representatives. Furthermore, according to the same memo, the Local and Professional Clubs had to elect their Coordinating Committees to participate in the election of their Prefectural Committees and Councils, to appoint the Regional Committee and finally to send delegates to the National Conference! In addition "ideologicopolitical zymosis" among the rank and file was strongly advised. All these were to be accomplished in a short period of time and under the supervision of appointed delegations from headquarters, who would have to guarantee and "report the unanimity to the Central Coordinating body.[26] Meanwhile the responsibility and the coordination of all these were left to the higher body of the organization which had been newly established by this memo: the Coordinating Body". This Coordinating Body, which was exclusively appointed by Papandreou himself was to be subdivided into a "broad" and a "narrow" section. The latter was to work closely with the President and meet every week, while the former was to meet "whenever the President considered it necessary".[27]

The promised goal of all this organizational activity was the National Conference. It was however a National Conference in which no one knew what was going to be discussed or what purpose was to be served, although Papandreou's memo, curiously enough, designated the participants. Thus in addition to the President, the participants to the promised National Conference were to be " delegates from the Local and Professional Clubs, the M.P.s, the founding members (of the Movement), members of the resistance and members of the previous Central Committees of the organization" – without even a hint as to how these last three categories were to be selected.[28]

It would be a rather fruitless exercise to try and categorize this intense organizational activity within PASOK according to any of the examples from the history of socialist parties. It is impossible to classify the numerous meetings, committees, sub-committees and councils according to any logic aimed at leading to the coherent structural development of the organization. The plethora of activities and mass involvement of the rank and file not only cancelled one another out but also kept the different tendencies of the Movement satisfied.

Thus, the *palaiokommatikoi* were happy both because they were given special consideration in the new organizational plans and because they could easily turn this organizational "industry", as practice soon proved it to be, into pre-electoral campaigns. The former PAK members and other radicals, due to their Leninist leanings saw in the plan some hidden intention for the creation of a well organized Movement. That part of the membership which still believed in the development of PASOK as a genuine mass democratically run, socialist party was also convinced, since the promised "National Conference" was supposed to be the form of a debate in which all the existing problems of the Movement would be solved. Thus once again the various currents of the Movement were able to see in the leadership's initiative for the organizational reconstruction of PASOK, the hope for the implementation of their own particular plan. This strengthened the leadership's role even more by further legitimizing its actions. In fact the latter became clear in the ease and the confidence with which the leadership managed to silence the occasional criticisms in the aftermath of the 1975 crisis,[29] a period during which no one could resign from the Movement without political cost to the leadership.

The other interesting characteristic of the post-1975 crisis in PASOK was the promotion of theoretical "debates". Theoretical or any other kind of discourse however, is not and cannot be, independent of its organizational framework. And furthermore in this case it was not independent of the "two speed" strategy which we described. Thus the undertaking of theoretical debates in the Movement contained all the controversial characteristics of the party's structure, as well as the new image it wanted to project for itself. First a detailed examination of the Movement's documents during the 1975-1977 period indicates that the theoretical debates were highly centralized. They were promoted mainly by the developing "nomenclature" of the leadership, particularly by Papandreou himself, and rarely by the rank and file. Secondly they were unfocused, creating further confusion for the ideologically weak portions of the membership. In any case there is no evidence that the latter ever discussed theoretical questions in any serious and structured manner. Thus in effect this attempt to promote "theoretical debates" was in practice a polemical exercise on the part of the leadership.

Furthermore, these theoretical debates were characterized by an ahistorical and polemical tone which further contributed, interestingly enough, to the building of the new moderate image, as the Movement attempted to cut itself off from all possible connections with the historical expressions of the marxist movement. For example the problematic around the Eurocommunist phenomenon which at the time was making European headlines was, in one of these theoretical proposals, boiled simplistically down to the politics of the Communist Party of Italy, the politics of which were in turn identified with traditional social democracy. As Papandreou put it, "the C.P. of Italy is the strongest social democratic party of Europe after the SPD".[30] In addition, Papandreou's attempt to deal with the relation between Marx and Lenin,[31] despite its theoretical accuracy and radical appearance, led, in a curious spin-off, to an out of hand rejection of the Soviet model of Communism, thus contributing further to the building of the new image attempted by the leadership. The *palaiokommatikoi* tendency of the movement and its constituency were able to interpret these ahistorical pronouncements freely (since they never really became the subject of internal debates) as evidence of PASOK's separation from the socialist tradition. At the same time, Papandreou's affirmation that "PASOK is indeed a Marxist non-dogmatic party"[32] was seen by the radical, marxist current of the Movement as an attempt to overcome the theoretical shortcomings of the dogmatism of the traditional left. But the latter, given the lack of debates over such issues within the Movement and the difference in the consituencies between the two tendencies (the radicals were composed only of a number of party activists, while the *palaiokommatikoi* had a larger appeal to the electorate) contributed further to the new moderate image of the Movement, becoming another piece in the strategy of the two speeds.

Overall however, in spite of the leadership directives in the 1975 period, the "theoretical debates" were far from becoming the major preoccupation of the membership. They were published in *Exormise*, PASOK's unofficial weekly newspaper and in the newly published *Meniaio Enemerotiko Deltio* (Monthly Information Bulletin) which was distributed only to the membership.[33]

As in the case of the structural question of PASOK, the theoretical controversies and confusion within the organization would not have grown to such enormous dimensions outside the context of the weak intellectual tradition of the Greek left. In addition, as we have previously noted, the theoretical and political sophistication, or lack thereof, of the organized left prevented it from providing PASOK's membership (at least the radical part) with a challenge and/or an alternative to the confusing vagueness dominating the Movement. The traditional left and the C.P.s in particular appeared to have great difficulty in dealing with their new partner/competitor on the left of the Greek political spectrum.

Two More Expulsions

The Trotskyists Forced Out

From the time of the Movement's establishment, Trotskyist groups and activists were accepted within the organization. As we have seen, their presence and activities were not a secret.

Some of them received key positions in its structure, not to the exclusion even of positions associated with the President's functions.[34] In fact Papandreou's special treatment of this tendency, along with his association with a Trotskyist group in his early days constituted the main reasons behind his much rumoured "Trotskyism". Although they were not accepted in organized groups, the majority of them belonged to two organizations and were faithful practitioners of "entrism".[35]

After the split in the summer of 1975 and given the intense attempt to reorganize the Movement, the Trotskyists became strong proponents of the "horizontal connection". This organizational plan was promoted by their members in the Organizing Bureau and implemented by others in rank and file clubs in which they had some influence. Through the opening of horizontal channels of communication within the Movement they aimed at bringing together the local and professional clubs without the mediation of the central collective bodies of the Movement. In this way it was thought that their influence in the ideological fermentation of the membership could be more effective. In fact part of this plan was already in action as some of the Trotskyist groups had been publishing their newspapers within PASOK since the spring of 1975.[36]

Trotskyist activity and influence in PASOK was limited to a very few clubs in the area of Athens. But through their strong theoretical background and their publications they started to give the Movement a rather militant and radical flavour. Having put their Leninism aside, they were not at odds with the initially loose political framework of PASOK. However, their discourse was definitely not in accordance with the new image of the Movement projected by the leadership. In the post-1975 period, their practice amounted to giving life to the original promise of PASOK for the legitimate existence of political tendencies in its structure. This was not only capable of undermining the leadership's attempt to control the organization, but it was also disruptive to the formation of the new image of the Movement. Implicitly the Trotskyist organizational proposition (i.e. development of "horizontal connections" among the local and professional clubs of the Movement) was to distance the rank and file from the leadership's control, with the hope of directing the clubs in which they had some presence closer to their ideas, contributing in this way to the left direction of PASOK.

Obviously, to the leadership, which was anything but keen to recognize the legitimate functioning of political currents within the organizaton, the Trotskyists' organizational practice was considered destructive factionalism and therefore had to be dispensed with. With admirable Machiavellian techniques, the leadership managed to build upon the uneasy relationship between the Trotskyists and PASOK B' and turn the latter against the former. The eventual result would in fact be the "legitimate" expulsion of the Trotskyist tendency from the Movement. Using methods almost reminiscent of Howard E. Hunt spy stories,[37] PASOK B', with the assistance of some of Papandreou's close associates, managed to prove the "factionist activity" of the Trotskyists. Thus in January 1976, a few days before the cancellation of the procedures for the meetings of the Regional Conferences and appointments of the Provisional Prefectural Committees, the editorial board of *Xekinema*, by an administrative order of the Executive Bureau, was expelled from the Movement's membership.

A number of clubs protested the decision, not so much out of sympathy for the Trotskyists but rather from their unspoken fear of the establishment of anti-democratic patterns in the settling of the Movement's internal differences. Once again the real issue i.e. the implications of the proposed "horizontal connection" were never discussed. The organizational dispute, in a crisis similar to that which occured in 1975, boiled down to a division between those who unquestionably supported the leadership's initiative for the initial expulsion, and those members who demanded, with various degrees of intensity, an explanation. Even the focus of the dispute itself and the overall basis of the dismissals were not clearly political, nor was the truth or falsity of the accusations of the "conspiracy of the entrists" even discussed. Rather the support and/or participation in the "horizontal connection" were taken to be sufficient grounds for dismissal. It was something which further underlined the intrinsic relationship between politics and organizatonal structure, of which PASOK can be seen as a prime example.

Although the Trotskyist crisis was not nearly as extensive as that of the summer of 1975, and the opposition was concentrated mainly in the area of Athens, some 300 to 400 members were expelled, and three Local Clubs of the organization were forced to resign. However, the Trotskyist appeal to the rank and file fell short of generating any real support.[38] The small number of Trotskyists in the Movement, and their traditional unpopularity both in and outside the organization prevented their expulsion from becoming a deep and prolonged crisis. In retrospect, the expulsion of the "Trotskyist element" was only a prelude to the upcoming and largely anticipated crisis with the Movement's youth. In contrast, that final crisis prior to the Movement's 1977 success, was to be resolved with the largest mass exodus which the organizaton would experience until the present day.

The 1976 Split

As we have indicated, the old membership of PAK, along with a number of other radicals, had for some time been identified, however vaguely, as the "left" tendency of PASOK. But the fact that this group had on several occasions been functioning as a united bloc, (as for example during the split with D.D.) should not be taken to mean that there were no differences within it. Part of this amorphous "left-wing" of PASOK identified the development of the Movement more or less with the political charisma of A. Papandreou and with his decisions – including that of co-operation with the *palaiokommatikoi*. Another part, composed mainly of the majority of the old PAK of the exterior, firmly believed that PAK should remain the backbone of PASOK. For some time this section believed that Papandreou himself was committed to this idea, something which likely explains their unconditional support of his initiatives. And finally another part of the "left-wing", although it had no historical links with PAK, was associated with it, since it was composed of the radical membership which had joined as a result of the organization's involvement in the mass movement (especially the students' movement). This group which had strong links with the Movement's students' section, appeared to believe in the development of PASOK, based on the political and ideological consciousness-raising of its membership, and its involvement in the mass movement.[39]

The differences between these factions were concealed for a long time and the PAK/radicals alliance remained united and supportive of the leadership's initiatives, especially the ones related to the organizational development of the Movement. However, according to at least two prominent members of the tendency,[40] as time passed, two things became clear: that Papandreou was not interested in any development of PASOK other than that which would lead to the expansion of its electoral influence; and that Papandreou was anything but committed to the idea of making PASOK a mass front of PAK. These realizations exacerbated the already existing differences within the group and polarized it into various sections *vis-a-vis* the leadership's initiatives. Thus, given the lack of any other established pattern of settling intra-party differences, the two new polarized factions set off on a long and fierce battle of slander, back-stabbing and administrative dismissals. It was a battle which tormented the organization for several months and was finally resolved after hundreds of the most dynamic activists of the organization had left the Movement.

Before we turn to the particulars of this latest split, let us examine the atmosphere developing within the organization, which was in fact the real factor generating the new wave of dissent. First of all, it was becoming more and more clear that the dominance of Papandreou over the organization was not simply a tactical move. The much advertised campaign to "revitalize the Local Clubs" was designed to revolve solely around Papandreou's speeches, allowing very little room for initiatives or free, independent, theoretical exchanges on the part of the rank and file.[41] Thirdly, the lack of well articulated politics made the attempts of the leadership to centralize any political initiatives of the rank and file seem a normal process. This meant that all the activities of the Local and Professional Clubs had to be reported and approved by the central authorities of the Movement, including initiatives which were exclusively related to minor local issues. Finally this situation was accompanied by an attempt to intimidate the membership and to boost its chauvinism. The members were told that they "should be aware of the consequences of possible mistakes" and that they "should not forget that our (PASOK's) enemies are waiting in the wings to injure (it) in order to weaken the main organized popular power of our country".[42]

In addition, it seemed that, with the exception of its increased concentration on the public relations front, the Movement lacked a concrete strategic plan. For example, in the sphere of international relations, the Movement created the appearance of being very active in meeting and promoting strong links with a number of parties and movements – without however discriminating between the politics of these organizations. Thus PASOK failed to make distinctions in the context of developing relations with such diverse organizations as the C.P. of Bulgaria, the Liberal Party of Norway, the C.P. of Cuba and the Ba'ath party of Syria.[43] The absence of any explicit concrete strategy, except of course the implicit electoralist one, in combination with the incoherent (at least at that time) proliferation of the Movement's Committees[44] and the controversial nature of some of its activities, were becoming more and more objectionable to PASOK's radicals. According to a prominent member of the radical tendency, the latter in combination with the growing annexation of the organization by "opportunists and all sorts of dubious political elements", the simultaneous strengthening of the parliamentary wing of the Movement,[45] and the parallel stagnation or rather drop in the active membership of PASOK[46] forced a good section of the PAK/radicals alliance to take action and try to extend its shrinking influence within the organization.

The main section of the Movement that provided the arena *par excellence* for the extension of the radicals' influence in the organization was the Movement's students' front PASP. For various other reasons in addition to being very much under the influence of the radical wing of the Movement, PASP appeared at that time to share the same kind of criticisms and concerns about PASOK's new direction. First PASP's intense day-to-day involvement in the mass movement was enriching the organization's collective consciousness as it was constantly encountering other left organizations. PASP's membership was exposed to marxist literature and to the history of the left which at the time still enjoyed great respect, at least among the radicalized section of the population. Secondly, a further common point of concern between the students' front of the Movement (PASP) and its radical wing was PASOK's inability to counter the C.P.s' opening of their collective processes with anything other than growing organizational confusion and a series of broken promises for democratic procedures in the Movement. These included: postponement in September 1975 of the promise made in June 1975 for the election of all collective bodies of the organization and appointment from above of "Local and Supervisory Committees and Provisional Organizing Echelons"; abolition of these Committees and appointment of a new "Organizing Bureau in November 1975; appointment of Provisional and Prefectural Committees; and the promise of a Panhellenic Conference [47] and the indefinite postponement of the latter by Papandreou in his speech at the Hilton Hotel to the membership in May 1976.

For these reasons, that part of the PAK/radicals alliance critical of Papandreou, thought they could use the Movement's student front as a starting point in their attempt to control the internal developments within the Movement. The plan of the functionaries associated with this group was the constitution of PASP as an autonomous organization with highly developed ideological and political positions within the framework of the Movement. They thought that this would not only increase student involvement in the mass movement but also, since students were participating in the Local Clubs of PASOK, easily radicalize PASOK and put an end to its growing moderation. Aware of the power of Papandreou both in and outside the organization, the group which was meeting under clandestine conditions, had decided that the plan was to be implemented without provoking a head-on collision with either the "President's clique" or Papandreou himself.

The first attempt to constitute PASP as an independent organization with a radical orientation had begun with the organizing of its First Panhellenic Conference in September 1975. As we have seen, this initiative was originally opposed by the official PASOK but finally the leadership had to give in since severe administrative measures were out of the question at a time when the wounds of the organization from the split earlier in 1975 were still fresh. Despite the limited preparation and brief proceedings, PASP called the conference the founding one of the organization, elected its first Central Council and called for the First Congress to be held the following spring.

Between PASP's conference and its First Congress (14 April 1976), the organization flourished, experiencing great popularity in the context of a Students Movement which was dominated by the left at that time. Many, if not the majority of the members of its Central Council, made connections with the PAK group, and their problematic, at least as it was crystallized in the documents of the organization's congress,[48] gave radical marxist interpretations to the premises of the 3rd of September. The changes in PASP, and its growing radical coherence soon started to be felt in PASOK as the new army of young radical activists which grew out of the student organization started to participate in the activities of the Local Clubs of the Movement.

From Movement of Protest to Leading Opposition Party

The plan for using PASP as the branch through which the "left" would make its presence felt in the Movement was working very well. Thus, when the First Congress of the organization came about, it had only to ratify and consolidate this strategy. With its new constitution, PASP was becoming an autonomous organization, of course within the overall framework of PASOK, but nonetheless independent of its "administrative interference".[49] The latter was objectively in contrast to the Movement's organizational course towards centralization, creating what led in fact to the first turbulence in the until that point harmonious relations between the PAK radicals who were working in close association with the youth and with PASP, and the leadership. Thus a few months after the congress, a prominent member of this group declared that PASP was under the influence of an organized faction (meaning PAK) with plans which would be disruptive to the Movement.[50] This "revelation" signaled the beginning of the administrative interference of the leadership which resulted in the ousting of the PAK/radicals from the Movement.

PASP's insistence on its autonomous organizational strategy clarified to the leadership that the only way in which it could stop the expansion of the PAK/radicals' influence within the organization was to control PASP. Thus, PASP became the arera in which the two factions, Papandreou and his inner-circle on the one hand and PAK on the other, were to resolve their differences. The leadership tried to transfer the live enthusiasm of its students' section from the autonomous PASP to the Youth Committee of the Movement. In the summer of 1976 the Executive Bureau appointed members of PASP to the Youth Committee of PASOK, hoping that the closer the problematic of the students was to the Movement's central apparatus, the less dangerous the former would be to the latter's orientation. The interference of the Youth Committee in PASP affairs and the transfer of members of the latter to the Movement created chaos in PASP. Fierce bickering started between the pro-leadership segment of PAK and other functionaries of the Center on the one hand, and the rest of the PAK radicals on the other, who in any case were the only ones in the students' front of PASOK who understood the implications of such initiatives. The huge mass of membership of PASP remained puzzled and sceptical. In the Second Session of the Central Council of PASP (12 September 1976) the debates which up to that point had admittedly been intense were replaced by slandering and fist fights.[51]

By the late fall of the same year, the hurricane marking the largest split of the Movement was visible on the horizon. Although largely silent and behind the scenes, the crisis was exacerbated even more when the most militant forces of PAK (especially those of Italy which controlled the entire PASOK organization in that country) showed no signs of compromise, and as more and more members of the rank and file were becoming aware of the leadership's apparent intentions – of co-opting at least the initiators of the autonomous PASP strategy.

On 18 October, A. Papandreou delivered a speech to a general meeting of the Movement's youth at which point the likelihood of the split became clear and basically only an attempt to minimize its political cost was really possible. Borrowing the techniques of the traditional communist parties, Papandreou came across fairly honestly and sincerely in his speech about the structural weaknesses of PASOK but he quickly attributed them to the membership's inexperience and to the existence of various tendencies in the organization. "There are", he argued "the *palaiokommatikoi*...the entrists (although they have been eliminated)...and a politically childish (tendency), which derives from revolutionary immaturity and ideological confusion..." In his attempt to further co-opt the rebellious membership, he praised the "ideological (and) political conflict within the organization...which can be seen as an element of political aliveness...but these antitheses can be overcome by the power of dialectics".[52]

Neither this direct appeal to the youth by the leader, nor the use of radical jargon and buzz words (such as dialectics, antitheses etc.), nor even the appointment of "leftists" to prominent positions[53] seemed to have any effect on the membership which had started to advance its criticisms.[54] Consequently, the leadership decided to take action and a few weeks later, nine activists of the Italian section of the Movement and members of PAK had (as PASOK's leadership again put it) "put themselves outside the Movement". In a Pan-Italian General Meeting of the Movement (11 December 1976), the membership unanimously decided to protest the decision and sent a five member delegation to explain their dissent to the President. The delegation was not even received in Athens, something which further fueled the protest of the Youth. A week later, in a Conference of the Movement's Youth in Athens, the Executive Secretariat of the Movement made allegations concerning the existence and functioning of organized political factions within the organization and announced the dismissal of another twenty members of PASP (ex-members of the old PAK of the exterior and those very much associated with it) who had allegedly been involved in those factions. At this point once again it should be stressed that neither Papandreou nor those in opposition to the leadership's interference talked explicitly about the organizational and theoretical issues involved in the dispute. The two sides were consumed, at least so far as can be ascertained from their written exchanges, with general complaints about, or justifications of, the situation within the organization and personalized accusations about procedure.

Nothing could stop the avalanche of the crisis. On 22 December, the secretary of the Movement's Youth G. Tzathas resigned in protest to the "continued decadence of the organization". "The decay of the organization," he argued "is evident in the continuous development of antagonism between an apparatus without ideology (and) principles, the apparatus of the central bodies of PASOK, and the rank and file which demands and waits in vain (for the development of a coherent strategy)...as well as in the development of the *palaiokommatikoi* network which grows stronger and imposes its choices upon the whole region".[55] Later the same night when members of the Central Council of PASP gathered for the Third Session of the Council in order to examine the situation, they found the headquarters of their organization occupied by a number of angry PASOK members which the leadership had mobilized from the Local Clubs and the headquarters nomenclature who managed by the use of physical force to keep the members of the Council out.[56] Fifteen out of the twenty members of the Council announced the complete autonomy of PASP and declared their will to continue the struggle within the students' movement.[57] At the same time the Disciplinary Council of PASOK in a brief meeting proceeded with the dismissal of another 48 activists of the Movement from both Greece and abroad. The dismantling of PASP and with it the dismissal of the majority of the PAK militants had become a fact. In the two months that followed, numerous members (especially from the youth and students' movements) and Clubs of the Movement were dismissed, resigned or became inactive. By the spring of 1977, when the crisis started to calm down, an estimated two thousand members of the Movement had left.[58]

For a short time it appeared that the crisis had seriously injured PASOK. Its front organization in the students' movement was in complete disarray, and its high school organization was dismantled. In some regions both in the countryside (e.g. Patra, Kavala, Ioannina) and abroad (e.g. Italy) it was left without local representation. But the leadership had succeeded in cleansing the Movement's ranks of the activists that were becoming (or could become) an obstacle to the building of its new image and its eventual development as an electorally strong organization. The membership's enthusiasm, which had been temporarily dampened, would soon be restored, and the Movement would be back on its course towards to the echelons of power. Besides the generally conducive political environment which made the possibility of PASOK's electoral success realistic, one of the major events which contributed to the revitalization of the membership's activism was the first Panhellenic Conference of the organization. However, in spite of the fact that in retrospect PASOK quite easily survived the 1976 crisis, the whole incident left deep scars upon the organizational practice of the Movement. From then on even the pre-existing theoretical possibility of political exchange and dialogue was to be replaced by simplistic polemics every time a new objection to the leadership's initiatives arose.

Crystallization of the New Political Orientation

The First Panhellenic Conference

After a series of postponements, the leadership announced the first National Conference of the Movement for the summer of 1977. The announcement was made at the peak of the crisis in PASP and although a national collective process was a long standing promise, it would be reasonable to assume that at least its timing was very much conditioned by the crisis and the overall internal criticisms concerning the lack of collective procedures in the organization. This latter anxiety of the rank and file was further clarified through the high degree of participation of the membership in the Regional conferences which preceded it and the eagerness of some of them in their questioning of the organizational strategy of the Movement, which was of course something that the recent crisis had brought to light once again. In fact some of them took the opportunity to openly criticize the leadership for its organizational practices up to that point. Many activists who criticized the leadership's proposals or initiatives were isolated or forced to resign and in any case none managed to become delegates or even to vote in the delegate selection meetings. Thus in the immediate pre-Conference period we once again witnessed a number of dismissals and resignations including the expulsion of entire Local Clubs which happened to exceed the limits of permitted criticism.

It is worth noting that as a consequence, PASOK walked into its first Panhellenic Conference with its internal field clear of any opposition, since forty out of the seventy-five members of its first (appointed) Central Committee and thirty-five of its second (elected) Central Committee were no longer members of the organization. In addition, only four or five members of the National Council of the radical PAK still remained in the Movement. Thus Papandreou's claim in his opening speech that "the Movement...in its two year life...had managed to delineate its ideological bounds"[59] came as no surprise to the critical observer of the Conference.

But as if the above described initiatives of the leadership during the two preceding years were not enough, Papandreou and the Executive appeared further to want to ensure control over the actual procedure of the Conference. The guidelines which were proposed and actually followed for the preparation and procedure of the Conference exposed the undemocratic nature of the whole enterprise. For example, Papandreou and the Executive "Organizing Committee" tried to control the participation of the youth, as it was this section of the Movement, being more steeped in a radical marxist problematic, which would likely be able to create problems and become on obstacle to the idea of "the opening of the Movement" which the leadership seemed to be promoting. Thus only one fourth of the delegates selected were permitted to be from the youth of the Movement.[60] In addition, and in accordance with the above mentioned guidelines, Papandreou and the Executive Bureau were able to take the floor as many times, and for as long as they wanted, while the delegates could only do so once and for only one and a half minutes. Furthermore, "the President can (could) remove any speaker (from) the floor if he is (was) talking off the topic of the agenda...(and) he has no right to talk on the same topic again...".[61] The special status of the President was further recognized as he was the only one who had the right to nominate delegates for the Central Committee while everyone else had to nominate himself if he wanted to run.[62]

Above all, the Conference crystallized the new direction which the leadership had chosen for the Movement. Both Papandreou's speech and the principles surrounding resolutions of the organization's constitution were the earmarks of this change. The references in Papandreou's speech to the unity of democratic forces which was the hot political topic of the day, indicated that the Movement was "...to fight with its doors open..." and that cooperation with the traditional left was essentially impossible since "they do not believe in the distinction between periphery and metropolis nor in the development of an alliance among the oppressed and underprivileged strata of the people for the achievement of our goals."[63] Given the absence of any specific class commitment on PASOK's part and in the context of the "over-parliamentarized" politics of the country, such a statement contributed to an exclusivist non-class definition of politics. In addition, this statement implied an off-handed long-term dismissal of the tactical alliance with the C.P.s on the basis of an alleged strategy which was anything but the truth. There was virtually no segment of the traditional left which was not operating on the basis of some form of dependency theory and which did not believe in, and promote, some kind of alliance between the "working class and its allies" (namely the peasantry and the petty-bourgeoisie). The acceptance of such statements by the membership was indicative of the PASOK rank and file's almost instinctual rejection of the traditional left, which Papandreou was using skillfully to both eliminate the possibility of any alliance with the rest of the left and to boost the membership's morale and enthusiasm as the general election was not far away.

Furthermore, the implicit and the explicit principles of the new constitution expressed at the Conference showed the new direction of the Movement, which was complementary to its organizational developments at that time. The Movement was to be organized on the basis "of the principle of Democratic Procedure which guarantees the characteristics of democracy and efficiency in the structure and the functioning of the organization."[64] The identification between democracy and efficiency had been timidly introduced during the pre-congress proceedings[65] but now had been elevated to a fundamental principle of the Movement.

However, the most noticeable new organizational principles to that effect were the further stimulation of the existing centralizing tendency and the elevation of the charismatic role of Papandreou to virtually that of an institution. The centralization of the Movement around the leadership was highlighted by the fact that fifty out of the sixty members of the Central Committee came from the list of Papandreou's nominations, and by the fact that two-thirds of its members were from the Athens region where the leadership had immediate and better contol of the organization.

Papandreou's role was not promoted to that of the power source of the Movement simply by default, but rather it was explicitly recognized as such. Thus according to the political proposal of the new constitution, PASOK was to be organized on the basis "...of a vertical expression of the rank and file..." as follows from the bottom up, i.e. from "the general meetings of the Local and Professional Clubs, the Prefectural Meetings, the Central Committee, the President and above all the Congress", which was to be called at some indefinite point in the future. In fact PASOK's organizational structure was anything but coherent. In any case, the explicit written organizational commitments of the Movement seemed to matter very little, as the organization became more and more centralized and under the more or less direct control of its leader.

The organization was in fact structured around three different levels of collective bodies: "Local level – First Degree Bodies (General Meetings of the Local and Professional Clubs and their Coordinating Committees); Prefectural level – Second Degree Bodies (Prefectural Meetings, Prefectural Committees and Prefectural Disciplinary Committees), Panhellenic (national) level – Third Degree Bodies (The Congress, the President, the Executive Bureau, the Disciplinary Council, the Committee for Financial Control)". Finally the constitution designated the creation of "Regional Committees" without defining any concrete jurisdiction or explicit position in the Movement's structure.[66]

At this point it is worth noting that the Conference did not vote on the Movement's Constitution, as the latter was conceived to be in the jurisdiction of the organization's promised congress and not of the present Conference. The delegates were only allowed to comment and note on the "Political Proposal of the Constitution" which, as we have said, defined the fundamental organizational principles of the Movement of "Democratic Procedure" as the "dialectic unity between Democracy and Efficiency".[67] These dialectics would guarantee the creation of a "vanguard-mass organization" which in turn "presupposes the organized presence (of the Movement) in every location where the people and the Youth live, work, entertain and educate themselves...(and) it means open methodology, comprehensive dialogue, united action in approaching the...people and not (an approach based on) abstract and incomprehensible ideological issues."[68] In any case the "Political Proposal of the Constitution" did not escape the generality and the spirit of these indicative statements and failed to make concrete suggestions as to how the organization was to move towards the implementation of these guidelines.

Furthermore, the special position of Papandreou was not simply recognized by officially and explicitly leaving the President's role outside of collective control. It was also the first time that his place in the Movement was rationalized on theoretical grounds. Thus in the Executive Secretariat's report on the activities of the various committees and bureaus of the organization we read that "the decisive parameters of the existence and the functioning of PASOK", (which is defined as "a unifying-organized MOVEMENT with a unified political entity) are the following: a) a uniform IDEOLOGY which gives it its political identity as a genuine mass movement which (can) promote a deep radical change of our society; b) its ORGANIZATION; c) its every day POLITICAL ACTIVITY by the people...; d) the personality of ANDREAS PAPANDREOU, who through his struggles and ideologico-political positions has established himself as the expression of the desires and the hopes of a wide social strata, for radical change".[69]

What is striking in these "paramaters" of the Movement's existence is that with the exception of "political activity", every other one has Papandreou himself clearly at its base. As we have seen, the basis of the ideological debates of the organization were Papandreou's speeches, while his position in the organization's hierarchy was second only to the Congress, which was to be called at some indefinite point in the future.[70] Even under these conditions, this political activity would not necessarily be an exaggeration if it was claimed that it took place largely to applaud the leadership's initiatives. Thus, statements such as this crystallized the implicit and silent dominance of Papandreou in the Movement and increasingly translated into the equation much feared by the organization's radicals – i.e. PASOK = Papandreou = PASOK. However, those most deeply opposed to it were now outside the organization, while those still inside started to rationalize it and indeed learned very quickly to love it or leave it.

Of course the tactics of promoting such a devastating organizational plan were not as crude as they might appear in the previous paragraph, but once again they were accompanied by both divisive and unifying proposals. In the aftermath of the Congress the leadership published a pamphlet which contained organizational directions to the rank and file with a distinctly radical – in fact a fashionably Maoist – flavour. Thus for example, the pamphlet on "Members' Duties" was full of advice about the need of the rank and file to engage in "self-criticism, criticism (and) rallying around" the decisions of the collective bodies of the Movement and on the need of the organization for "raising consciousness".[71] But such a document can reasonably be seen as an expression of the "two-speed" tactic, since its limited circulation and the lack of an organizational structure which would allow the implementation of its premises, amounted to a guarantee of limited impact.

To sum, PASOK's First Conference crystallized and institutionalized the dominance of Papandreou over the Movement and the already existing tendency towards centralization. From now on, with virtually no turbulence, especially since the aroma of electoral success was penetrating the senses of the membership, the organization would bend to the leadership's wills and initiatives with unprecedented ease. The Movement's initial promise for participatory democratic collective internal processes would be permanently buried under Papandreou's omnipresent weight. The members would be called into collective action – both internal and external – only when all possible opposition had been neutralized and they were ready to applaud, ratify and actively execute the leadership's political plans. In other words, the Congress served as the necessary and complementary expression, at the organizational level, of the moderate image which the leadership was building in order to prepare itself for the next election.

Both the weakening of the government's popularity, as well as the drastic reduction in the Center's political appeal, appeared as golden opportunities for Papandreou's party to capitalize upon. But neither the appeal of a charismatic leader nor the politics of moderation were sufficient to win an election. Electoral successes are more often the result of the workings of an appropriate and well oiled mass electoral apparatus. Having fostered the undisputed charisma of a leader with national appeal and having created both politically and structurally a moderate image for itself, PASOK now had to turn its attention to its electoral apparatus in order to ensure electoral success.

Building the Electoral Machine

The undermining of the rank and file within the organization with regard to its active participation in policy formation, should not be taken to mean that PASOK's leadership was not interested in further recruitment or that the membership was about to abandon its political involvement in the mass movement. After the last split, the Movement's leadership had achieved both a reasonable degree of political uniformity and established its control over the party's structure. Consequently with the upcoming election in sight, it launched a strong campaign of mass recruitment. As we have seen previously, PASOK had made recruitment drives customarily following the membership losses of its internal crises. But now in the post Conference period, especially in the context of the Movement's new organizational development, the recruitment drive was purely electoral in nature.

Although the beginnings of this type of recruitment and electoral organizational strategy were evident several months before the Conference, it was after the summer of 1977, theoretically at least a good year-and-a-half before the officially anticipated date of the election, that the organization intensified this practice. The recruitment drive was put forward as a strategy for the "qualitative and quantitative development of the Movement". However, from the actual organizational proposal and memos, and particularly from the actual practice followed, the emphasis was on "quantity" and not on "quality". This emphasis was expressed through the leadership's appeals to "open up the organization...(to) all social classes and strata,[72] so that the Movement will become "representative of all people both in terms of age and occupation".[73] The latter advice was accompanied by explicit restrictions on Youth participation in the Local Clubs.[74] Given the radical nature, the high level of political energy and the relatively high level of political sophistication of the Youth, the latter point can be interpreted as a lack of interest on the part of the leadership to truly upgrade the political education of the membership and to promote rank and file participation in the policy making of the organization.

PASOK's recruitment drive was met with great popular enthusiasm, as Local Clubs of the Movement mushroomed throughout the country. The success was particularly vivid in the countryside where even in the most remote villages, the Movement started to make its presence felt with the opening of local offices. Putting aside the reservations one might have in terms of the qualitative basis of this organizational expansion, in the context of modern Greek politics it was probably one of the most significant developments in post-war Greece. Organized local structures besides the *kommatarches* had been absent in the countryside which made active political participation very difficult. In addition, active political affiliation with the parties of the opposition, thanks to the watershed of the civil war, was considered hazardous as oppositional political activism was often identified with traitorous activity. In this context, PASOK's new organizational structure must be seen as a considerable positive contribution towards the legitimization of such political participation and activism.

The newly established Local Clubs were however, far from being places in which their constituents could expect to develop politically. The clubs suffered from an absence of concrete assigned political projects which often meant that they became meeting places for coffee and backgammon. In addition, these Local Clubs were frequently dominated by the *palaiokommatikoi*. However the lethargy on the part of the Local Clubs was replaced during electoral campaigns (national or municipal) by intense activism and effective mobilization (as the 1977 election displayed) of the local votes.

While the Movement's regional organizational development was pointing in the direction of the creation of an electoral political machine, its activity in the student and trade union movements appeared to be of a different kind. As we have said, PASOK had developed ancillary organizations, which although they were integral parts of the Movement, were used as front organizations for its involvement in the respective mass movements. Through PASP (Panhellenic Militant Students' Front) and PASKE (Panhellenic Syndicalist Movement of Greece), PASOK managed to make great advances. In the February 1977 student election PASP won 18% of the vote[75] and by the beginning of the same year, PASKE controlled 28% of the Executives of the white collar trade unions and associations.[76]

Despite initial appearances, PASOK's involvement in the mass movements should not be understood as a radical departure from its overall electoral orientation. These auxiliary organizations were under the close supervision of the leadership. In addition, the overall political practice in the Greek mass movements was one in which partisanship dominated and political autonomy was the exception. Thus these front organizations were merely the transmitting belts of the political choices and the orientation of the party as they were defined by the leadership. In addition, their electoral involvement within the various organizations of the mass movement were good training for the national campaigns. And furthermore, the electoral successes of these front organizations were at least symbolically conducive to PASOK's attempt to expand its appeal as the new national election started to become visible on the horizon.

To sum, PASOK entered the electoral race with its ranks free of disturbing radical elements, and with a structure highly centralized around its charismatic leader. Its initial promise of "democratic procedure" had not only failed to become the living guide of the organization but it was even dropped from the Movement's fourfold slogans without in fact the slightest reaction on the part of the membership.[77] In addition, as in its electoral programme and overall discourse, even the previously descriptive references to classes had been replaced by the general classless juxtaposition of the "privileged" versus the "underprivileged", the goal of "social liberation" had been concretized as the promise of "equality of the citizens in front of the law" and the "socialist transformation of Greek society" had become a mere "social change".[78] That is to say, PASOK had fully developed for itself the image of a serious, non-threatening, political alternative to the right. Thus the Movement and its leadership faced the elections with great optimism. The electoral results were in fact to confirm this optimism.

A Change in Image

Generally speaking, it is quite natural for a political party to want to maximize its electoral popularity and influence. PASOK was about to carry this rule to great lengths. In the 1977 election PASOK entered the campaign with a political platform which was quite removed from the 1974 politics of the Movement. This of course does not mean that the new politics, as they were expressed in both Papandreou's speeches during the campaign and "the Guidelines for Governmental Policies", were completely foreign to its "Declaration of the 3rd of September". Rather it meant that the new politics had given a particular twist to the "3rd of September", which strongly suggested the new moderate direction of the Movement.

In the introduction to "The Guidelines for Governmental Policies" PASOK claimed itself to be the "political carrier of change and socialist transformation of the Greek society". But it was quick to point out that "change does not mean revolutionary overthrow (although the Movement had never implied so in the past)...but simply the acceptance of the goals that have matured in the popular consiousness and their pursuit with proper planning and prudence according to and to the extent to which the objective potentials of the Greek economy and society allow".[79] In addition to this introductory statement, there are overall two main characteristics of the programme which highlight PASOK's new direction towards moderation and strongly suggest that the radical dynamism of the Movement had been removed as a realistic possibility. The actual theses proposed were not much different from those in the opening Declaration of the Movement. However, in a fashion very much reminiscent of a social democratic approach, the country's problems were discussed in the absence of their structural causes and as a result, the new political statements of the Movement concerning tactical questions and issues had been essentially disconnected from any strategic goals. Secondly, in contrast to the populist character of the "3rd of September", the new programme had a strong technocratic flavour. Its reference to two major issues, namely foreign policy and the economy can be seen as good examples of PASOK's new politics.

"Independent foreign policy and national independence" still remained PASOK's priority as can be seen from the structure of its electoral programme (constituting the first section of the document) as well as from the overall understanding of the country's politics along dependency theory lines. Thus PASOK's electoral programme stated that "the basic goal of the policies of the Movement is (was) the abolition of (the country's) dependence upon foreign centres of decision making i.e. the freeing of the country from the foreign political and economic alliance, complete withdrawal from NATO and the dismantling of the foreign bases".[80] The Movement based its foreign policy on the assumption that Greece belongs simultaneously to Europe, the Balkans and the Mediterranean and that its goal is (was) to work for the creation of a "Federated Socialist Europe". PASOK laid great weight on the Mediterranean dimension of the country's existence, since a common position among the countries of the Mediterranean was seen as necessary for the "confrontation of the international capitalist institutions and negotiations with the capitalist industrialized countries within the framework of North-South dialogue".[81]

Taking these policies at face value, it is difficult to detect any great departure from 1974, since it does fall within the framework of the "3rd of September". However, there is a clear difference in analysis and the context in which the Movement's proposals on foreign policy were put forward. In the founding document of the Movement, the goal of "national independence" and the overall foreign policy of the Movement were conceived as closely related to the organization's internal policies and the pursuit of "social liberation". In fact "national independence (one of the four principles of the Movement) was understood as the prerequisite of the goal of "social liberation" and vice versa. In 1974, foreign policy was seen as having "decisive importance for the country's future...since it decisively influences the conditions and the possiblities of our internal development".[82] Such an analysis opened up the possibility of a class based discourse/practice and, as we have seen, allowed, at least theoretically, the radical wing of the organization to pursue it. PASOK's 1977 electoral programme however, "failed" to make or even to imply such a connection. Now "national independence" was seen along exclusively nationalist lines and in juxtaposition to the orientation of the Karamanlis government's foreign policy, which rested essentially on the dictum "We belong to the West".[83] Thus although the Movement's new positions on foreign policy were nominally not much different from the ones stated or implied in the first three years of its life, the way in which they were argued had radically changed, indicating an overall shift of the organization towards moderation and more conventional politics.

Furthermore, on the controversial issue of the country's entrance to the EEC, we can clearly detect a similar trend of departure from the 1974 positions. Instead of the previous outright rejection of the idea of Greece's entrance to the EEC, the 1977 electoral programme stated "PASOK still believes that our (Greece's) participation in it (the EEC) would consolidate the peripheral and satellite role of the country and make national planning impossible...(therefore)...the right solution is not annexation but the development of relations with the EEC within the framework of a new special agreement (of the Norwegian type), which will allow Greek control of the national economy as well as the movement of capital and goods".[84] In addition, PASOK now considered the issue a very serious one that "no government should handle without the mandate of the Greek people, which can be given only through a genuine referendum".[85] This new position on such a sensitive issue might not seem to the sympathetic observer a radical shift from its original policy. However, it is rather hard in this context not to sense a certain amount of electoral opportunism as the Movement struggled to present itself as a thoughtful, serious and realistic political alternative.

But if PASOK's new foreign policies were only by implication a departure from its founding positions, its economic policies, as they appeared in its 1977 electoral programme marked an explicit abandonment of the Movement's initial radicalism. The programme euphemistically referred to the Movement's economic policies as "the realization of economic democracy". However, the implied radicalism of this title bore no relation to the content of the subsequent pages. PASOK's original criticism of "the free market system which does not, and could not, produce commodities according to the social interest" and its strategy of "eliminating the exploitation of man by man" through "extensive socializations of productive economic sectors...under a general decentralized planned economy in which workers' participation is guaranteed,[86] was completely forgotten. Instead it was replaced by the "prime pursuit...of securing the prerequisites for a self-suffient national development".[87]

This meant nothing other than "increasing the rates of growth", "the rightful coordination of investments for the capital equipment of various sectors" (and) the "reduction in income inequalities both among regions and among social strata and the security of a minimum standard of living for all Greeks".[88] For the "protection of these policies of change" PASOK did "keep" its original promise to socialize the means of production. However, this time the reference to "socializations" was confined to specific sectors which were already state owned or controlled, i.e. the financial, transportation and energy sectors.

In spite of the Movement's continued efforts to remain vague, PASOK did not manage to escape some striking contradictions, indicative of its contradictory project to both continue and expand its political influence among the popular classes, as well as to legitimize its existence within the echelons of the power bloc. Thus the original hostility to capital and capitalist choices (e.g. the "down with capital" slogans of the 1974 electoral campaign or even the hostility to the free market economy) had been replaced by the affirmation that "private initiative...appropriately supported and oriented within the framework of the national programme and freed from the suffocating presence of monopoly capital will become the basic lever for the country's development".[89]

In other words, PASOK had abandoned its overall hostility towards capital in general in favour of a vague opposition to monopoly capital. Of course monopoly capital was anything but threatening to Greek capital, especially in the secondary sector. But PASOK even tried to qualify this anti-monopoly sentiment which at least by implication referred to foreign capital, by promising that it would "continue pursuing foreign capital investments but in such a way that they would not prohibit the development of domestic business initiatives". And having said that, it was quick to point out that "it will for the latter (i.e. domestic business) create the conditions for it to be competitive...with its foreign counterpart. To conclude its controversial economic proposals, and in an effort not to leave out the petty-bourgeoisie, PASOK promised that special consideration was to be given to "medium to small businesses and to the cottage industry".[90]

In addition to these striking changes in the Movement's 1974 and 1977 political platforms, one could not help but notice a distinctly technocratic flavour in the proposed policies. The Movement's economic plan and its complementary policies (e.g. industrial, agricultural, transport, tourism polices) were to be implemented under the auspices and the scrutiny of the state which was to promote the education of and studies by "a large number of technicians and scientists...for the study (and) the application of modern technology" in the solution of various problems. Even the remnants of the Movements's initial populist gestures for participation and democratic control of the country's economic planning, which were now referred to as "democratic planning", were presented in a complicated fashion with the creation of numerous committees, subcommittees and councils so that even to the sophisticated reader they appeared as incomprehensible and remote possibilities.[91] One gets the same kind of feeling from the attempted application of the original promise of "self-management", which was redefined as political self-management. The latter was in turn confused with the need for decentralization and the overall upgrading of the civil service,[92] and therefore ultimately was lost.

The Electoral Result

The electoral results indicated that PASOK was changing the political map of the country more rapidly than even its most optimistic supporters could have imagined. Papandreou's Movement had almost doubled its electoral support (25.33%) and had increased its holdings in parliament sixfold, gaining 93 seats in the 300 member parliament, relative to its only 15 seats in the previous one. This had set the trend and generated the widespread belief that the Movement was well on its way to completing its short march to power. This becomes more clear when one looks closely at the electoral result. PASOK was changing the political map of the country and doing so in a most radical and universal way. Judging from the vote distribution,[93] its support did not come only from the urban centres and selected provinces with long liberal democratic traditions (i.e. Crete and Achaia), but from all constituencies in all regions of the country. Furthermore, the Movement's support in the urban centres was not only limited to the downtown constituencies but came equally from the predominantly working class suburbs. Although unfortunately data do not exist which would allow a more detailed class breakdown of the PASOK vote, it is reasonable to assume that in the 1977 election the Movement built a firm foundation for becoming a truly national party with support which both broke with regional traditions and crossed class barriers.

New Democracy's support had, on the other hand, diminished, from its previous 54.37% to 41.85%, i.e. from its 219 seats to 172. Although Karamanlis' party undoubtedly showed signs of decay, the overall losses of the right (both extreme and moderate) were not as dramatic, since the National Front (an extreme right wing party) gained 6.82% and 5 seats. The big loser of the election was the Center, which had fallen to 11.95%, and only 15 seats. This was largely a result of the fact that the Center was no longer the only anti-right, non-communist political force. It seemed that PASOK's success was to become the Center's "kiss of death" as the former's electoral gains came primarily from the withering away of the Center. The C.P., which managed to dominate the communist left, had gained a respectable 9.36% and 5 seats. The C.P (Interior) took part in the "Alliance of Progessive and Democratic Forces" which just barely managed to win 2.72% and only 2 seats, indicating that the former had lost the hegemony of the communist left to the C.P. and that the Eurocommunist banner would have a hard time flying in Greece.

There were many reasons for the unprecedented advance on PASOK's part. To the critical observer, PASOK's own account of the result reads like a history of the Movement's shift towards moderation. The first reason given by PASOK for its success was the transformation of the organization, especially after December 1976 (the time of the dismissal of PAK) from "closed theoretical clubs into live political shells". To the leadership this transformation meant "the cleaning up of the strategy and the tactics of the Movement (and) the development of its organization into a close relationship with the popular masses".[94] Of course having followed the Movement's development fairly closely, we know that this "cleaning up" amounted in effect to the silencing of the membership's opposition (especially its radical portion) through one method or other, tight leadership control, and the overall moderation of the organization's image.

Furthermore, PASOK saw its success as a result of its intense activism highlighted by "the missions and the presence of the Movement's functionaries in the country-side".[95] In other words, it pointed to the exact organizational pattern that the left had strongly objected to because it was leading the Movement towards an uncontrollable electoralism and at least by implication to a very strong focus upon parliamentarism. In fact the leadership made no pretence concerning the importance of the latter since it listed the effectiveness of its parliamentary presence as one of the key factors in the Movement's electoral success.[96] These may well be realistic appraisals of the organization's "strengths", but the Movement went even further and in a sense became cynical. In a "fit" of self-awareness, it underlined the role of Papandreou and even implied an identification of the Movement with his charisma. Thus, the post-electoral organization finally appeared to have become aware of Papandreou's dominance in the Movement, and having passed through the period of objection, then of toleration and consensus, it seems to have ended up loving that dominance. Consequently, over and above the President's parliamentarian and public relations abilities, his personality was recognized as one of the most significant factors in PASOK's electoral effectiveness. "He spoke frankly on the positions of PASOK and...did not hide anything from the people...this was the truth, the honesty, the commitment (the) persuasion of PASOK."[97]

Furthermore, PASOK identified as strengths those tendencies which in foregoing analysis we had identified as vices. Thus its "Guidelines for Governmental Policies" identified the organization's electoral advantages on the basis of the fact that they were the product of "the study of experts over many months...", not because of their political orientation. In other words, PASOK's appraisal of its electoral popularity identified the advantages of its electoral programme not on the basis of the content of its proposed policies but on the quality of the researchers involved in its production. In fact, many observers within the country, particularly journalists, had identified this technocratic tendency of PASOK and had interpreted it as a turn to the right. Of course however, neither the Movement nor Papandreou himself were willing to openly accept something like that, except by implication. Answering a question on the subject, Papandreou himself did not deny that these changes in the policies and the overall image of PASOK had taken place but that "...this should be considered a realistic adjustment of the general principles (of the Movement) to (respond to) the concrete situation."[98] Thus PASOK's politics were now understood not as points in mobilizing support for implementing the strategy of the transformation of society but rather as proof that PASOK was a serious enough party to manage the country's problems.

In addition, the leadership did not fail to include the contribution of its electoralist strategy of the previous two years on the list of items in PASOK's self appraisal. It clearly recognized the benefits of focusing the organization's activity almost exclusively upon electoral expansionism, of the Movement's emphasis on parliament and the elevation of the latter into a political arena at least equivalent to if not more important than the struggle outside the House. The latter was a perception that not even the main-stream membership had, at least up to that point, dared to put forward so clearly. Thus in effect PASOK was recognizing the precalculative nature of its new moderate orientation.

Finally, PASOK appeared to be well aware of the appeal of democratic structures and made no mistake in promoting its "democratic functioning" as one of the positive components of its electoral success. It also became apparent that at least the leadership's perception of a democratically functioning party was rather formalistic, serving one and only one purpose – the expansion of electoral support. Thus, the Movement's "democratic functioning" was to be "proven" from the "official election of the Central Committee and the Executive Bureau" which in effect destroyed the "myth of the personalized party" and contributed to the organization's electoral triumph.[99]

In spite of the contradictions in PASOK's analysis of the electoral result (e.g. the simultaneous underlining of the importance of "democratic functioning" along with the recognition of the extraordinary role of Papandreou), it was in fact quite correct and did constitute a faithful descriptive statement of the Movement's overall electoral strategy. However, what PASOK failed to recognize was the "contributions" of the other parties to its electoral success. No one can tango solo, and politics, particularly elections, are no exception. Papandreou's Movement would not have been so successful if it had not been for the state of affairs within the other parties and particularly those of the opposition.

The most obvious contributor to PASOK's electoral growth was the withering away of the Union of the Democratic Centre (E.DE.K.). Despite their unity and electoral experience, the forces at the centre of the Greek political spectrum had shrunk significantly, emerging from the polls with only eight percentage points, and in 1977, they lost 45 of their previous 60 seats in parliament. The "Centre" had been closely associated with the tradition of the mid-sixties radicalism. However, despite its efforts, which can be highlighted by its main-stream[100] discourse, the Union of the Democratic Centre neither managed to adjust to the radicalized post-dictatorship political environment within the country, nor to become a distinct and separate political entity from the renovated "New Democracy". Consequently, it was permanently left behind, and PASOK with its charismatic leader and an appropriately radicalized discourse was left virtually on its own to capitalize on the situation.

A further contribution to PASOK's electoral advance was made by the traditional left, particularly the C.P. which appeared to be uninspiring to an electorate with a strong petty-bourgeois flavour and a long anti-communist tradition. The C.P.'s unrefined, vague and grandiose strategy for the formation of an anti-imperialist "New Democracy"[101] and its obvious lack of participatory democratic patterns could only guarantee it a limited success – i.e. its triumph over the C.P. of the interior for the dominance of the communist vote.

Similar sorts of inadequacies were clear in the ranks of the "other" C.P.(Interior). This time however, the problem did not derive from a vague ultra "revolutionism" but rather from a remarkably reformist practice, (for a radical, let alone a communist party)[102] in its stubborn effort to mechanistically apply a Spanish flavoured Eurocommunist strategy. In the 1977 elections, the C.P.(Interior) chose to join the "Alliance of Progressive and Left-Wing Forces" along with some independent leftist groups, radical Christians and social democrats.[103] Although the "Alliance's" electoral platform was undoubtedly much more radical than the C.P. of the Interior's discourse had ever been, it was not convincing, since due to its relative size, the party dominated the "Alliance". Furthermore, by appearing on the eve of the elections, the "Alliance" had a hard time changing the label imposed on it by its critics as an "opportunistic electoral scheme".[104]

These were the conditions and the dynamics of PASOK's

competitors which constituted in effect one of the most significant factors in its unprecedented success. PASOK did not of course admit this publicly. But to us, it was these peculiarities of the parties of the opposition that enabled PASOK to achieve its position as the leading opposition party in the 1977 election and to enter the election which followed with a real chance of success. The Movement and its leadership appeared to be well aware of the dynamic of this situation and they did not fail, immediately after the election, to set the psychological mood in preparation for its coming to power. "The People" PASOK's leadership stated, "with their vote on the 20th of November laid the foundations for change...They made the first but basic step in the great march for the fullfilment of (their) desires...(therefore) now (our) responsibilities are even larger and the work which is waiting for us even more difficult."[105]

There is no doubt, at least in retrospect, that this type of analysis and understanding of the electoral result on PASOK's part set the stage for the further development of the Movement. However, this did not imply a reorientation of the basis of this development. The basis and the priorities of PASOK's development had been established in the pre-election period when its leadership won *carte blanche* control over the organization, and when activity was almost exclusively confined to electoral and parliamentary politics. In effect, there was to be no change in PASOK's strategic orientation, except that more of the same initiatives were to be added in response to the fact that the organization was only a step away from forming the next government.

In the post 1977 period the Movement's development was to be marked by the consolidation of the power of its parliamentary wing which had just risen victorious from the election and was becoming the *raison d'etre* of the Movement's political activities and existence.[106] Furthermore there was a strengthening of the technocratic tendency as the Movement fell victim to its undemocratic structures and increasingly recognized the awesome task that would be involved in actually running the country. The result of all these was the immediate creation of an electoral campaign climate not only within the Movement but also in the country's political arena. This lengthy electoral campaign started the night of the 1977 election and ended four years later with a Pasokean triumph.

Notes

1. *Ideologikes Arches* (Ideological Principles). Athens: New Democracy, September 1974.
2. Ibid.
3. It appeared that the nationalizations, in addition to being the result of the Government's economic planning (nationalization of a number of banks as the Government wanted to establish a firm state of control in the financial sector) were politically motivated. For reasons which did not go beyond personality conflicts, the Government nationalized enterprises whose capital was the result of state protectionism during the military regime. The nationalization of the Andreades conglomerate is a case in point.
4. Greece, due to its traditionally close economic ties with the Arab countries, was one of the very few countries in the world which had not recognized Israel until the mid-1970s. In the post *metapolitefse* years, Greece's voting behavior in the UN, especially on the issues related to the Near East, was closer to the non-aligned nations than that adopted by Western Europe. Kahler Beare, *Political Forces in Spain, Greece, and Portugal*. London: Butterworth Scientific, 1982.
5. In effect, it was only during the 1974-1977 period that the Government of New Democracy started to respond to the issue. In fact, it was not until 1979-80 that working hours were reduced from 48 to 40 per week.
6. The Government fell short in the introduction of actual reforms, however it initiated study committees on the issue, which at least by implication, recognized the legitimate nature of the issue. It was an important development in the context of the rather traditional Greek society.
7. Commission of the European Communities, "Annual Economic Report 1983 – 1984",quoted in *The Greek Economy in Figures 1984*, Athens: Electra Press Publications, 1984.
8. Ibid.
9. After a decline of 4% in real wages in 1974, there was an increase of 6% in 1975, of 9% in 1976, and 9.2% in 1977. Source: *European Economy*, 1983 quoted in *The Greek Economy Today*, published by the Center of Planning and Economic Research, Athens, January 1984.
10. R. Macridis, *E Ellenike Politike sto Stavrodromi* (Greek Politics at the Crossroad), op. cit., p. 83.
11. *The Greek Economy in Figures 1984*. op. cit.; and *Concise Statistical Yearbook of Greece*. Athens: National Statistical Service of Greece, 1981.
12. See for example the militancy displayed during the strikes in the

From Movement of Protest to Leading Opposition Party

Ladopoulos papermill (14-20 August 1976 and mid-September and October of the same year); the series of farmers mobilizations over prices during which tractors were repeatedly driven into downtown Athens or to the nearest urban center and complete chaos was subsequently created (Athens fruit producers 21 June 1975; tomato producers August 1975; cotton producers 14 November 1975) as well as the militancy in the public sector in D.E.E. (national hydro company), in the Telephone company and by teachers. These mobilizations were indicative not only of the centrality of the Greek state in the societal conflict, as most of the protest was almost exclusively concentrated upon the government but also of the "classless" overpoliticization which characterizes the Greek social formation as we have shown in the first chapter of this work.

13. See below in this chapter.
14. See for example Papandreou's interview "Socialismos sten Ellada, Pos kai Pote?" (Socialism in Greece, How and When?) in *Ta Nea* 3 November 1975; and A. Papandreou "E Mesogiake Kene Agora" (Mediterranean Common Market) in *Eleftherotypia*, 2 February 1976.
15. *Exormise*. 19 September 1975.
16. Answering a question in court during the trial of the Junta, I. Alevras claimed that he was a firm fighter for democracy since "...when the communist-gang came to Mount-Parnitha we put forward our bodies to stop them." (Summer 1975)
17. Papademetriou M.P.: "...our ideal is Scandinavian socialism". Dailies mid-August 1975.
18. *Exormise*. 9 May 1975; and a press conference 3 March 1977.
19. Letter of A. Papandreou to the leaders of the other parties on "Democratic Unity" 9 November 1976, published in *Miniaio Enemerotiko Deltio* (Monthly Information Bulletin – M.I.B.) November 1976.
20. During the geological research of the Turkish boat "Hora" (Summer 1976) Papandreou proposed a dynamic confrontation of this "challenge" amounting a declaration of war against Turkey ("Sink the 'Hora'"). After severe criticism from all parts of the political spectrum however, he qualified his statement. Interview in *Exormise*. 13 August 1976; also reprinted in the Movement's publication *Apele sto Aegeo* (The Threat in the Aegean) Athens 1977. pp. 85-92.
21. In addition to going to great lengths to claim that in his opinion the army was always in the service of the nation and that only a small gang of traitors was responsible for the coup (trial of Junta Summer 1975), Papandreou went so far as to claim that the army officers had nothing to do with the government's change of the Junta's death sentence to life imprisonment. In fact he went so far as to claim that "... on the contrary the armed forces wanted to severely penalize the guilty parties..." (speech in Patra 11 October 1975). This type of gesture,

which consciously separated the role of the army from the social and economic structure of the country became a permanent trait of PASOK's politics as we will see below.

22. A focus on the parliamentary definition of political practice is of course infused by the system and the practice of the other political parties. However PASOK made no effort to break out of this conventional pattern. It was in fact strengthening this pattern. See for example how *Exormise* announced the leaders' debates in parliament: "Tomorrow the battle in parliament" (16 October 1975) while a week earlier, rather faithful to the two speed pattern of PASOK, an editorial article was explaining how "The first signals from parliament proved the dated nature of personalized parties" (10 October 1975).

23. From a letter to the President and the Executive Bureau from the "Committee of the Expelled and Administratively Removed Members of PASOK", 5 October 1976, published in *Sosialistike Poria*, 30 November 1976. The same feeling was also expressed in both the meeting of PASP in Salonnika and in the A' Panhellenic Conference of PASP(see next note).

24. The Youth section of the Athens organization took the initiative to call the A' Panhellenic Conference of PASP in September of that year. The initiative was initially opposed by the Executive Bureau, Papandreou, and the nomenclature of the central offices, since the initiative would have given PASP an autonomy and a separate (unwanted) dynamic which would have been "dangerous" to the image of the Movement as a whole. Finally however, the strong voices of the Youth sections of Salonnica, Patras, Italy, England and partially Germany supported the initiative and given that at the time a split without major political cost for the organization was impossible, the Students of PASOK proceeded with their first collective function. (Interview with D. Tzouvanos *op. cit.* and T. Demetre, "E Diaspase tou '76 sto PASOK" ("The Split of '76 in PASOK") *Fylladio* Vol. 6, August-October 1983.) A similar situation developed in Salonnica's PASP when the membership bypassed the by this time vocal objections of the "Center" and organized a regional general meeting of the organization and took a very critical stand vis-a-vis the overall condition of the organization despite the presence and police-like interference of a delegation from the Athens headquarters (6 October 1975).

25. Memo No.11, 18 July 1975 from A. Papandreou to the Movement's organizations.

26. Ibid.

27. Ibid.

28. Ibid.

29. In the general meeting of PASP in Salonnika (see above), A. Tsohatzopoulos, member of the Executive Bureau and very close to Andreas, responding to criticisms on the organizational confusion of

the Movement, declared that "The Movement's structures are given and unalienable (and) no protest can change the situation." Quoted in *Sosialistike Poria*, 18 October 1985.

30. Speech in Corfu, 9 August 1975. Athens dailies. Published in a pamphlet by PASOK's Committee of Publications in 1978.

31. A. Papandreou, "O Marx, o Lenin kai e Dicatoria tou Proletariatou" (Marx, Lenin, and the Dictatorship of the Proletariat). *Exormise*. November 1975.

32. Ibid.

33. As a further case in point regarding the shallow nature of the theoretical activity of the party at the time, it is worth noting that this bulletin was originally published by the Center of Research and Enlightenment, as a monthly communication instrument within the Movement in which the rank and file could put forward its problematic. However, it soon turned into a publication in which the speeches and the articles of the leader were reprinted, or the orders of the Executive Secretariat and other Bureaus were communicated to the organization's clubs. This peculiarity determined the frequency of the publication, which was anything but regular.

34. See for example the role of Paublist Livieratos who worked in the Executive Bureau and the Center of Research and Enlightenment.

35. The two major Trotskyist organizations were the "Workers' Group" and "Workers' Democracy", which after the mediation of Ted Grand of the British Militant Tendency, were unified.

36. These were *Xekinema* (The Beginning), published by the "Workers' Group", Mayday 1975 and *Deltio* (Bulletin), published by "Workers' Democracy", September 1975.

37. From many sources we have clear indications that a strong network for spying on the movements and the activities of the Trotskyists was established. See R. Elefteriou "PASOK: Apo ten Auto-organose sten Graphiokratia" ("PASOK: From Self-organization to Bureaucracy" *Anti* Vol. 241, 16 September 1983; also the details given by one of "the president's men", K. Laliotes, in a speech delivered to the Athens membership of the organization (30 August 1976) and published in *Monthly Information Bulletin* October 1976.

38. The Trotskyist appeal had no further effect on the Movement. Finally its hard core and some sympathetic ex-PASOK members formed the "Political Link of the 3rd of September" (October 13, 1976), continued to publish the *Xekinema* and continued to maintain some "semi legitimate" relations with the official PASOK, enjoying a marginal political existence on the fringes of the Movement.

39. Interview with Tzouvanos. op. cit. and T. Demetre; also "E Diaspase tou 76 sto PASOK" (The Split of 76 in PASOK) in op. cit.

40. Interview with Tzouvanos. Ibid. and T. Demetre, Ibid.

41. Memo No. 14, 6 June 1976, in *M.I.B.* June 1976, p. 11.

42. Ibid. p. 20. The attitudes frequently expressed through the leadership's references to the historical mission of PASOK should be kept in mind because, in retrospect, they contribute to an explanation of the membership's arrogance in the post 1981 period.

43. See "Anaphora Doulias tou Graphiou Diethnon Schesseon" (Report of the Bureau of International Relations) in *M.I.B.* April 1976.

44. See for example, the Committees' reports in Ibid., May 1976.

45. Interview with D. Tzouvanos. op. cit.

46. According to an interview with 13 members of PASP who were kicked out of the organization (8 January 1977), the number of active members of PASOK was "according to the most optimistic estimates only 7000," while according to A. Papas, a prominent member of the International Relations committee who resigned "based on the internal reports of the Organizing Bureau," the active membership was not more than 1500-2000." A. Papas, "Gramma Papaiteses" (Letter of Resignation) in *Sosialistike Poria* 15 December 1976. Unfortunately, no official accounts of the Movement's 1976 membership are available.

47. Memo No. 12. 28 January 1976 in *M.I.B.* April 1976.

48. *Dokumenta tou A' Panelleneou Synedriou tes PASP* (Documents of the A' Panhellenic Congress of PASP). Athens: 1976.

49. Ibid.

50. It was S. Tzoumakas, (later a cabinet minister) a member of the resistance and very much associated with the youth who communicated this to the leadership. T. Demetre, op. cit. and interview with D. Tzouvanos, op. cit.

51. T. Demetre. op. cit. and D. Tzouvanos. op. cit.

52. *M.I.B.* November 1976, pp. 3-12.

53. Papandreou appointed K. Laliotes, a prominent young member of the Movement and a member of PASP who was considered to be on the left of the Movement to the Executive Bureau. Assuming his new position, Laliotes distanced himself more and more from PASP and the radical tendency, and became one of Papandreou's closest associates and later a member of cabinet.

54. See for example the highly critical "Syzetese tou Kinematos kai tes Anaptixes Katastatikou, tou PASOK Italias" (Discussion of the Movement and the Development of the Constitution of PASOK of Italy) published in a series of four articles in the weekly *Exormise* between 26 November and 23 December 1976.

55. G. Tzathas, "Gramma Paraiteses" (Letter of Resignation) in *Sosialistike Poria*. 25 December 1976.

56. In fact, the events at PASP's headquarters that night were dark indeed, with violence implied as pieces of broken furniture and chains were waved in the air by the leadership's supporters.

57. PASP Press Conference, 8 January 1977.

58. T. Demetre, op. cit. and D. Tzouvanos, op. cit.
59. *Exormise.* 16 July 1977.
60. "E Diadikasia tes Ekloges ton Antiprosopon kai tes Synelefses" (The Procedure of the Preceedings of the Delegates' Appointments and the Conference). *M.I.B.*. May 1977, pp. 21-37.
61. R. Eleftheriou, "Ennea Chronia PASOK" (Nine Years of PASOK) *Anti*, Volume 242, 30 September 1983. p. 27.
62. *M.I.B.*, op. cit.
63. "Politike Esegise" (Political Proposal) by the President and the Executive Secretariat in *M.I.B.* May 1977, pp. 3-22; and in *Sosialistike Poria*, 16-30th of July 1977.
64. "Politike Esegise tou Karastatikou" (Political Proposal of the Constitution) in *M.I.B.* May 1977.
65. "E. Diadikasia tes Ekloges ton Antiprosopon kai tes Synelefses" (The Procedure of the Preceedings of the Delegates' Appointments and the Conference). op. cit.
66. "Politike Esegise to Katastotikou" (Political Proposal of the Constitution). op. cit.
67. "Politike Esegise" (Political Proposal) by the President and the Executive Secretariat in *M.I.B.*. May 1977, p. 7.
67. Ibid.
69. *M.I.B.* July-August 1977, p. 15. Capitals in the original.
70. At this point we must note that the First Congress of PASOK was called only in May 1984.
71. *Ta Kathikonta tou Melous tou PASOK: E Katasktese tes Sosialistikes Synedeses* (The Duties of PASOK's Members: The Conquest of Socialist Consciousness). PASOK Publications 1977.
72. "Apophase tes Organotikes Epitopes tes Syndiaskepses yia tous tropous Anigmatos tes Organoses" (Decision of the Organizational Committee of the Conference on the Ways of Working Towards the Opening up of the Organization) in *M.I.B.*, February-March, 1977.
73. Ibid.
74. "Apophase tes Organotikes Epitropes tes Syndiaskepses yia tou tropous Anigmatos tes Organoses" (Decision of the Organizational Committee of the Conference on the Ways of Working Towards the Opening up of the Organization). op. cit.
75. This established PASOK as the close second power in the Greek Universities. The C.P. was the strongest and the rest of the left parties and groupings controlled a commanding 60%.
76. The same numbers were 14% for the C.P., 16% for the C.P. of the Interior, 5% for Sosialistike Poria, 13% for E.DE.K., 1% for E.D.A., 6% for the independent left, while the right controlled only 17%. Report of the period June 1975 – December 1976 by PASOK's Committee of the Professional organizations of Scientists and Artists *M.I.B.*, op. cit. January 1977.

77. It was at that time, before the 1977 election that the Movement's weekly *Exormise* scrapped the slogan from its front page.

78. Ibid. It is worth noting that the term "socialism" was completely absent from its political verbal discourse – we cannot of course be certain of all the electoral speeches but this was definitely true so far its programme was concerned. This is not the basis of our claims concerning PASOK's new direction since it is our conviction that a political party in a particular context can practice politics which genuinely aim at a structural and radical change of the society. However, in PASOK's case, as we hope was clarified in this chapter, the absence of "socialism" was the byproduct and the culminating symbolic expression of all the developments in its structure as well in the peculiar Greek social formation, which led to its divorce from the possibility of following a radical/socialist course.

79. *PASOK Katefthidiries Grammes Kyvernitikes Politikes* (PASOK Guidelines for Government Policies). *Exormise*. Special Edition, no date available. Introduction.

80. Ibid.

81. Ibid.

82. *Te Thelie to PASOK* (What does PASOK Want). Athens: KE.ME.DIA., 1974, p. 25. also see Appendix I.

83. *PASOK Katefthidiries Grammes Kyvernitikes Politikes* (PASOK Guidelines for Government Policies). op. cit. "We belong to the West" was K. Karamanlis' summation of his government's foreign policy orientation, which became a slogan for the right and a focus of criticism for the opposition.

84. Ibid.

85. Ibid. It is worth noting that PASOK's proposal to have a referendum on the question of the EEC was borrowed from the C.P.'s programme.

86. *Te Thelie to PASOK* (What does PASOK Want). op. cit., pp. 15-16.

87. *PASOK Katefthidiries Grammes Kyverniktikes Politikes* (PASOK Guidelines for Government Policies). op. cit. Emphasis in the original.

88. Ibid.

89. Ibid.

90. Ibid.

92. See for example the proposal for the creation of "Developmental committees of the Base" as part of the "Democratic Planning". Ibid.

92. Ibid. See section under title "Measures for Political Self Management."

93. M. Drettakis. *Vouleftikes Ekloges* (Parliamentary Elections). Athens: 1982 and "Results of the National and European Parliamentary Elections Since 1974." in *Athens News Agency*. 8 May 1985.

94. Editorial in *Exormise*. 25 November 1977. Similar views were expressed in many post electoral analyses of PASOK, but we quote from this one because it was the most comprehensive.

95. Ibid.

96. Ibid.

97. Ibid.

98. Interview with A. Papandreou in *To Vema*, 26 February 1978. Also published as a pamphlet *A. Papandreou Synentefxe tes 26 Flevary, 1978* (A. Papandreou Interview of 26 February 1978). Athens: PASOK Publications Committee, No. 13, 1978.

99. *Exormise*. op. cit.

100. See its electoral programme: *E.DE.K.: Manifesto 77* (E.DE.K.: Manifesto 77) Ellinikes Democratikes Ekdoses, October 1977.

101. *K.K.E: Eklogiko Programma 1977* (C.P. of Greece: Electoral Program 1977). No exact date of publication.

102. For example the party operating under the strategy of "National Anti-Dictatoric Democratic Unity," the infamous E.A.D.E., in its effort to fight the ghost of the junta, tried to form coalitions with the right in various mass movements including the trade union movement.

103. The organizations which formed the "Alliance", besides the C.P. (Interior) were: E.D.A. – the remainder of the old front organization with a very moderate independent leftist orientation; *Sosialistike Protovoulia* (Socialist Initiative) – a small organization of political personalities with a clearly traditional social-democratic orientation; *Christianike Demokratia* – a small organization of radical, anti-right Christians; and finally *Sosialistike Poria* – the outgrowth of the PASOK's 1975 split which had managed to attract a good number of, and the sympathies of most of the rest of, the independent marxist left.

104. This claim was at first rather unfair since the "Alliance" was a product of the initiatives of *Sosialistike Poria* to form an anti-right political coalition which had started a year and a half before the election. However, after its mediocre showing in the election (2.7% of the vote and 2 seats), the C.P. (Interior), the most powerful partner, decided to withdraw its membership and in effect break the scheme, thus fueling the criticisms.

105. *Exormise*. 25 November 1977.

106. For a symbolic expression of the fact that the Movement's organization was taking a back seat to its parliamentary wing, we can site Papandreou's call for the meetings first of the Parliamentary Caucus (for 2 December) and then of this Central Committee (3 December) just two days after the election.

5

The Short March to Power Completed 1977 – 1981

PASOK had planned, worked, predicted and "achieved its electoral goal"[1] when it became the country's leading opposition party. There is no doubt that the result of the November 20th election was "the Movement's triumph" or rather proof that its leadership's strategic choice of a short cut to power was working. But above all, as we have said, PASOK's unprecedented achievement had radically changed the political map of the country and with it the course of its history. In the years that followed immediately thereafter, we were to witness PASOK's preparation for entering state power. This was highlighted by the gradual but rapidly increasing entrance of technocratic elements into the Movement, as well as the definite abandonment of the discourse/practice which was still remotely related to the possibility of a socialist project.

Thus, as one one might expect, there were no radical changes made in PASOK's structure after 1977. With the exception of some fine tuning of the Movement's organization which allowed it to respond to, and capitalize on, the shift in the electorate, there were no organizational developments of the kind which we witnessed in the previous period. The splits and the factionalism which had characterized the Movement's development up to this point, were absent in the period now under consideration.

This absence of deep internal turbulence in the organization should not be taken to mean that there was no opposition to Papandreou's court from the remnants of the Movement's initially promising radicals, or that there were not numerous members who fled or were forced to leave the organization. In fact, more people left the organization in the 1977-1981 period than ever before. However this time the dismissals and/or departures remained isolated, factionalized and without effect.[2] Obviously the expulsion of the "organized" founding tendencies from the organization and the new centralized structure of the Movement had not helped to unite the voices of the opposition. Furthermore, it soon became clear that the sweet taste of electoral "victory" and more importantly, the realistic prospect of forming the next government could be nothing but destructive in any oppositional efforts to gain even minimum support and to become effective.

In the following pages, we will examine the Movement's preparation for its final assault aimed at ousting the right from power, as well as the political context within which this preparation took place.

PASOK Becomes the Unchallenged 'Vehicle of *Allage*'

The Movement, as we saw in the previous chapter, was well aware of its potential, and having been successful up to that point was not about to change its ways. This was reconfirmed in the 2nd session of its Central Committee which the organization held just three months after the election (i.e. on 23 February 1978). The "...avoidance of any sectarian move" in order "to express all the masses which are thirsty for the big change"[3] became the main guide to the Movement's further initiatives during this period.

The 1977 election had proven PASOK the winning team on the Greek political spectrum. This, in combination with the withering away of the Center, made the Movement the pole of attraction to numerous politicians and functionaries of the parties of the Center. However, this was not perceived by the leadership as alienating to its ideological and political positions, since it fell under the general attempt of the Movement to promote the "entrance of the masses".[4] While PASOK seemed capable of rationalizing the opening of the organization's gates to the center by deliberately disregarding the distinction between consciously aware functionaries and "ideologically trapped masses", it was quick to call the few members of the center who chose to join "New Democracy" apostates.[5] There was no doubt that the Movement knew its success depended on its claim over the center of the political spectrum, which the decline of the parties of the Center had left open.

Construction of the Moderate Image Continues

To PASOK, claiming the political centre did not mean the abandonment of all the radical positions of the Movement, but rather their mild modification through the simultaneous introduction and maintenance of a "new" moderate discourse. Thus, PASOK maintained its fierce attack and radical jargon against the US and NATO.[6] However, during this period all the issues related to "national liberation" (e.g. the Turkish claims over the Aegean, the Cypriot issue) were seen almost exclusively as a conspiracy of the Americans and the CIA. This emphatic thrust in the Movement's confrontation of foreign policy questions, not only cut off the possibility of connecting issues of "dependency" with the domestic structural problems of the country, something which, as we saw, was present in the 1977 electoral programme, but most importantly it implied that a mere change in the way things were administered could (and would) change the situation. In fact, now the problems of the country's foreign policies, as they were symbolically summarized by Karamanlis' infamous dictum "we belong to the West" were merely perceived to be "unfortunate mistakes of the past which have not yet been revised".[7]

Almost as if this new understanding of foreign policy were not sufficiently "innovative", Papandreou did not fail to surprise even his sharpest critics with some shocking and politically rather dangerous statements. In his effort to reserve for himself and his party a supra-patriotic place in country's politics, explicitly militarist and chauvinist positions were aired by Papandreou and his party. For example Papandreou, in a 1978 interview in the Movement's paper, claimed that "if we do not acquire tactical nuclear weapons of our own, we will become Turkey's satellite"![8]

Similar patterns of change in the Movement's policies can be seen on the burning issue of the country's membership in the EEC. As we have seen, by proposing a referendum, PASOK, in its 1977 electoral programme had already removed itself from its previous outright rejection of the government's plan to join the EEC. However, even the tone of this part of its opposition was now reduced, when it came to be expressed, not as part of the Movement's general strategy for the "socialist transformation of Greek society", but rather in the context of concrete technical problems which the imminent ratification of Greece's membership would have caused.[9] A close examination of this new twist in the Movement's position on this question, can reasonably well support the argument that the Movement, facing the prospect of forming the next government, was trying to build bridges with that part of capital which would have to make the most significant adjustments after the country's entrance to the EEC. Through such efforts, the Movement would likely gain further support as a serious, legitimate political alternative.

In a fashion similar to that of the pre-1977 period, the expression of the Movement's radicalism was taking place more within the framework of the organization than publicly, in this way faithfully continuing the "two speed" strategy. Thus often both in *Exormise* and in *M.I.B.*, we witnessed the publication of theoretical articles, which although they often had a strong polemical flavour, were far more radical than the public discourse/practice of the organization overall. Classic examples of this type of radicalism were the polemical rejection of social democracy and the continuation of a populist criticism of what the Movement called "the establishment". In the first case, social democracy was perceived to be the mere voice of the US; therefore it was unacceptable because it did not comply with the Movement's narrowly nationalistic attitude,[10] and because it was determined out of hand to be irreconcilable with PASOK's 3rd world brand of politics.[11] In the second case PASOK's radicalism was exhausted in personified attacks against the "established class", a notion which was widely used but was in fact far removed from any substantive notion of class. This evil "establishment" was criticized either on the basis of tax evasion,[12] or because individuals held more than one position and salary, or even because they drove a particular model of Mercedes that "the average Greek cannot even dream of driving".[13]

The initiatives leading to the further moderation of the Movement's politics became more and more vivid as the time of the next election approached and the organization's leadership became more eager to gain its public legitimacy through moderation. Thus, side by side with an anthropomorphic attack against economic imperialism [e.g. "Foreign capital ravenously eats up the (country's) state banks"],[14] was the affirmation that "not only does PASOK not plan to discourage private initiative but on the contrary it plans to encourage it...(with) a new policy of incentives and...public investment",[15] while next to the latter, the promise for the socialization of some major economic sectors was advertised.

Controversy concerning the Movement's radicalism went beyond the specific themes we have mentioned, and embraced all the Movement's references to, and attempts at, theoretical analyses. Of course this does not mean an abandonment of the "two speed" strategy, as the peaceful coexistence of both a radical and a moderate aspect of PASOK's theoretical discourse continued. In fact, this pattern continued to be the main source of the organization's strength as it attempted (quite successfully) to keep its quite diverse political components in line. However, some initiatives of the strategy during this period, reached vividly contradictory levels. For example, pictures of Galbraith and Marx magically appeared as complementary on the same page of *Exormise*.[16]

Parliamentarianism Consolidated

This idiosyncratic road to moderation was pursued even further through the continuation of the emphasis on the role and the importance of parliament and the Movement's parliamentary caucus.[17] This characteristic had been developed, as we have said, well before this period, but in the pre-1981 life of the Movement it became a more conscious effort. It became more rationalized by the organization at that time since the party's nomenclature was trying to ground it theoretically. These theoretical attempts to rationalize the importance of parliament were taken on by some recognized leftists of the organization who located this "parliamentarianism" at the low level of the people's political consciousness. They thus explained the organization's concentration on parliamentarianism by the fact that the masses had no other developed forms of conducting politics.[18]

The 1977-81 emphasis on parliament must be seen as qualitatively different from the creeping parliamentarianism of the period which led to the 1977 election. In the previous period, the parliamentary activity of the Movement had gradually come to be seen as being as important as its extra parliamentary activity. However, in the 1977-81 period, the organization had completely reversed the order of its priorities when PASOK went so far as to conceive of all the other political activity of the organization as a mere extension of the activity of the parliamentary caucus. In the resolutions of the 5th session of the Central Committee, extra-parliamentary activities were no longer considered the basis of the Movement's politics or even (as in the pre-1977 period) an equal partner in the activity of the parliamentary caucus. Now parliamentary activity has explicitly become the center of the Movement's activity, while "extra-parliamentary activity is (simply) an extension of PASOK's positions in parliament".[19]

The impact and significance of PASOK's new perception of parliamentary activity and of politics in general were not limited to the political orientation of the organization. Political parties contribute to the definition of politics as much as they are conditioned by the political environment. PASOK's new definition of politics was no exception, particularly since day by day the Movement was extending its popularity. A number of events in Greece's recent political history had undermined the legitimacy of parliament, at least as the exclusive arena of conducting politics. These included: the wheeling and dealing in the parliament of the mid 1960s which had resulted in the dissolution of the government of the enormously popular Center Union (with 53% of the popular vote); the series of scandals and biases associated with almost every electoral competition prior to that; the failure of parliament to stop the colonels' tanks from rolling into downtown Athens in 1967; as well as the radicalism which followed the long night of dictatorship. Furthermore, in the immediate post – *metapolitefse* era, the Greek people had indicated through intense extra-parliamentary activity that there was more to politics than that which was conducted within the walls of the House, or which took place around the election of "national representatives". Consequently, PASOK's new definition of politics, as it derived from its political discourse/practice, must be seen as an important contributing factor to the reorientation and narrowing down of the popular definition of politics.

To be fair however, it would not have been possible for PASOK to so dramatically change its political priorities if the other political parties had not been engaged in similar discourse. Of course, so far as the "bourgeois parties" are concerned, the emphasis on parliamentary politics was not surprising. What was really interesting, and in fact conducive to both PASOK's change in political orientation and the overall parliamentary redefinition of Greek politics, was the similar political practice in which the two C.P.s were engaging. On the one hand the C.P. had made reaching 17% of the popular vote [20] the exclusive goal of its political practice, while on the other hand the weaker and therefore less ambitious C.P. of the Interior centered its political activity on the parliamentary survival of the party.

Technocracy Consolidated

The most important change in the definition of politics to which PASOK contributed was not the emphasis upon parliamentarianism, but rather the technocratic attitude which it managed to diffuse into the country's political discourse. The previously latent idea that the management of, and solutions to, the country's problems were merely a matter of the rational application of expert proposals, now penetrated all aspects of the Movement's existence in the period under examination. Of course in the context of Greek politics, and particularly given the widespread, what we might call "idealistic voluntarism", the inherent realism, or even cynicism, of the technocrats could be considered rather positive. However, in the absence of a mass, democratically organized movement and in combination with a parliamentary definition of political activity, this technocratic focus could only function in opposition to the dynamism of the popular movement.

The growth of the technocratic tendency in PASOK was to be seen primarily in its debate on the economic issues of the country to the extent that it was difficult to distinguish any politics apart from those implied in their technocratically "objective" analysis. Papandreou's previous opposition to the government, based on the Movement's principle of "social liberation" was now replaced by criticism of the government for its lack of "...a consistent and integrated developmental policy".[21] Futhermore, the Movement's principle of "national independence" in the context of economic discussions was translated into complaints about "...the passive dependency of Greek industry...which is not able to mobilize local capital (or to propel) the capitalist role of the state".[22] The rise and the strengthening of technocratic ideas in PASOK were also occasionally translated into puzzling statements which were strikingly alien to the Movement's alleged radicalism. For example, it was this technocratic rationale which led PASOK to vote with the government on a new marine labour legislation. The anti-labour nature of the legislation was beyond any doubt, as it was transferring security payments from the employers to the workers and further contributed to the dismal state of labour organization in that sector, but PASOK overlooked all these as the "efficiency and international competitiveness of our marine fleet" appeared more important.[23]

As one might have expected, this change in PASOK's politics opened the gates of the organization to hordes of technocrats who literally inflated the Movement. During the period under examination the Movement's structure experienced an invasion of numerous highly educated members, usually with foreign graduate degrees.[24] This influx was also actively encouraged by the Movement's leadership which organized a number of conferences inviting many such people. For example, under the auspicies of Papandreou himself, PASOK organized a conference (in Corfu, 23-30 August 1979) of Greek economists who worked abroad, to discuss the economic problems of the country and the Movement's response to them. PASOK also organized an "international scientific seminar" on "The Transition to Socialism" (in Athens, 30 June to 3 July 1980) and on the eve of the election, it sponsored another "symposium" on the theme "Greece and Socialism", (in Athens, 23-27 September 1981). As can be seen from the published documents of the meetings, they lacked any coherence or focus,[25] however, they contributed significantly to building the image of a serious party, whose politics were scientifically correct.

In spite of their political inexperience, the technocrats were to become the third element, along with the remnants of the radical tendency (Marxists) and the all powerful *palaikommatikoi*, in the delicate balance between the various political currents of the Movement, and a very significant factor in the future dynamic of the organization. The Movement's deliberate initiative to open its doors to technocrats should be seen as a realistic response to what we might call its "fear of power". As time passed and the distance between PASOK and power shortened, and as "the moment of truth" in which the Movement would have to implement its promises came nearer, a diffuse anxiety became one of the main traits of the organization. Many sources indicate that the organization was confronted with this pressing reality,[26] and given that in practice it had rejected the challenging path of democratic participatory organizational development, it had no other choice but to rely on technocratic rationality. Finally this technocratic turn of events was naturally reflected in the Movement's organizational structure, in which the complexity of countless committees and bureaus became even more convoluted, and the "Governmental Programme" of the Movement became the exclusive affair of these technocrats.

As election time approached, the presence of this technocratic attitude of PASOK became more and more intense, as it was combined with the existing characteristics of the Movement, giving them a new twist and contributing further to the Movement's image of being a party capable of becoming a future government. Thus for example, there was an effort to have the prospective prime minister seen as not only having the charismatic qualities of his predecessor, but also a command of all technical aspects of the country's problems.[27] Obviously, Papandreou's academic background assisted in this expansion of his image. This technocratic capacity was particularly significant since it put Papandreou on top of the rapidly growing technocratic current within the party. In addition, the combination of the all powerful charismatic personality of Papandreou with the virtually unlimited capabilities of the technocrats not only worked to give Papandreou an objective advantage over the other leaders of the political spectrum, but it also led some to attribute almost omnipotent capacities to the organization.

Finally, this technocratic orientation of the Movement started to overtake its day to day criticisms of the government's policies, as it gradually started to change the basis of its opposition to the government. Originally PASOK had based its criticisms on the principles of its founding document and then moved on to personalized its attacks on the government. Now however, New Democracy was criticized simply for "its inability to introduce the proper measures to overcome the country's problems".[28] In this context, PASOK's understanding of the country's problems was more and more framed in technical and economistic terms (e.g. income inequalities and deficits as merely the results of extensive tax evasion).[29] Finally, its analyses were accompanied by the promise that PASOK had both the will and the ability (thanks to the Movements's expertise) to handle these problems.

The Final Fine Tuning

In addition to the consolidation of PASOK's parliamentary and technocratic orientation, it appeared that as the time of the election approached, its leadership realized that some further initiatives would be necessary if the Movement were to be recognized as the unchallenged "vehicle of *allage*" (change) and if it were to replace New Democracy as the "nation's party". These initiatives were often symbolic in nature and aimed at the fine-tuning of the overall image of the organization before the election. Thus they were intended to further expand the Movement's base, and to reinforce the Movement's image of pragmatism and seriousness in order to legitimize PASOK as the next "party of the nation". It was necessary for it to be acknowledged as such, not only domestically but internationally, and it was also important for the Movement to give further guarantees of the smoothness of the imminent transition to power.

This change in PASOK's politics and overall image had obviously one aim – the electoral victory of the Movement in the upcoming elections. More than anything else, winning elections ultimately boils down to numbers, which means the avoidance of any exclusive political affiliation with a specific social class or stratum, and the embracing of the widest possible social support. Of course from the very beginning, PASOK had no exclusive class affiliations. Its reference to social classes, as we have seen, was a descriptive list of social classes, groups, and professions (i.e. "PASOK is the Movement of the working men, the foremen, the youth, the self-employed, the pensioners,...") which in combination with other aspects of its 1974 programmes left some room for interpretations along the lines of class analysis.

The pre-1977 distinction between "privileged" and "underprivileged" was a step away from even the possibility of class analysis, as even the descriptive reference to the working class was taken over by the virtually all-encompassing "under-privileged". In the pre-1981 period, in its effort to extend its social base and even further remove itself from class references, PASOK appeared with a new and innovative term (for the Greek political vocabulary: the "lower middle strata" (*micromesea stromata*). The realization of its newly introduced strategy of the "democratic road for a socialist *allage*...is (was) impossible without the participation of the lower-middle strata.[30] The theoretizations on the role of these strata as the "backbone of the *Allage*" did not take place in the context of the problematic about the rise and the role of the new social strata or in the context of the traditional question of the left forming social coalitions. In fact, PASOK did not even attempt to define them, leaving in this manner, strong implications that just about everyone could belong to this magic category. Thus, it appeared that PASOK's special consideration of the lower-middle strata was due to their large number and the consequent electoral weight which they carried.[31]

Another area in which the Movement did some fine tuning in order to further improve its already moderate image and become more acceptable to the population at large (but especially to the power bloc) was in giving guarantees for the "smoothness of (the) change" which it advocated. These guarantees were given both through explicit statements as well as through some symbolic initiatives. As the election approached, the Movement's leadership became almost obsessed with the idea of proving that PASOK's victory was not going to generate any disruption in the nation's fortunes. In many editorial and first page articles in the Movement's press, it was promised that "PASOK (was) the guarantee of smoothness and change".[32] However, since *verba volen* (words fly) the more effective initiatives in this field were not these statements but rather some of the Movement's active political positions with concrete material substance. Thus while PASOK supported most of the militant trade union activity during the period when it was directed against the government, it was at this time that the trade union front organization of the Movement (PASKE) displayed quite different behaviour in its field activity. PASKE declared a militant strike of the marine enginers "improper" and confronted it accordingly.[33] Exceptions such as this, did send signals concerning the limits of PASOK's radicalism. In retrospect, these incidents can be seen as a comforting signal to capital in terms of the smoothness of the transition and underlined PASOK's capacity to restrain the trade union movement if and when this was necessary.

As we have mentioned previously, the Greek social formation and consequently, the Greek political culture are characterized by what we have called a Janus-like syndrome. The Greek political culture is nationalist (with occasionally alarmingly chauvinistic overtones), and internationalist at exactly the same time. Thus to conduct politics in a legitimate and effective fashion in the Greek context, political leadership must adopt some dimension of diffused nationalist political discourse, but at the same time it must gain international acknowledgement and prestige either by promoting good public relations with western heads of state, and/or by using a foreign country as an ideal model.[34]

In the months immediately prior to the election, PASOK appeared to be actively aware of this need, and more than ever before worked to further its leader's legitimacy and international prestige. Thus during this period, in order to build "PASOK's glamour and prestige",[35] Papandreou traveled probably more than he had ever done. And while he was visiting the capitals of Western Europe and the Middle East, the Movement's publications were consistently trying to publicize his visits and receptions as achievements of his internationally acknowledged personality, thus further boosting his status as a national leader.[36] The essence and the material substance of these trips were rarely (if ever) explored, as emphasis was given exclusively to their public relations aspect.

Furthermore, PASOK appeared to be giving importance not only to international public relations but also, in accordance with the established left wing pattern, it tried to affiliate itself with the international experience of other socialist parties. It was in this context that PASOK took the initiative, which was very much a symbolic by-product of the moderate turn of events within its organization, to organize meetings among the leaders of the Southern European socialist parties in order to coordinate a socialist strategy for the South.[37] This would not have been so unusual if PASOK had not previously criticized these parties as traitorous to the people and as "instruments of US foreign policy". There is no doubt that the Socialist Party of France (PSF) had a rather radical outlook at the time and there were many positive things which could be said concerning its record as a government up to that point.[38] However, the European parties involved had done very little which could justify this positive change in PASOK's attitudes. Thus, the Movement's initiatives *vis-a-vis* the "socialists of the south" should likely be seen simply in the context of boosting the organization's international prestige.

Another important part of the fine tuning which the organization underwent was its further effort to elevate Papandreou into a national figure. Even more than before the Movement continued to give tremendous importance to its leader's contacts with Karamanlis. Just as described in the previous chapter, even routine contacts between the two men were seen by the organization as an opportunity to further prove that its leader had the prestige and the legitimacy to become the leader of the new government after the election. There were even strong rumours in the country at the time that Karamanlis himself, after his departure for the presidential post (spring 1980), following the dramatic changes in PASOK's politics and given the disarray in "New Democracy", had started to view Papandreou very favourably as the next prime minister.[39] When by 1981 the organization ceased to refer to its leader as "comrade president" and referred to him instead as the "leader of the official opposition" the Movement was not only boosting this much desired very tame image for Papandreou but was also symbolically highlighting the changes which had taken place in the Movement as a whole.

The changes in the previously radical premises of the Movement, the entrance of technocrats into the organization, and the continuous emphasis on nationalism as PASOK's leadership chose to indicate through gestures to the armed forces,[40] were not only a telling story about the Movement's "pragmatism" but were also creating the impression that the Movement was turning to the right. This latter view was reinforced by Papandreou's not unusual appearances at social gatherings of the Greek bourgeoisie. The left, and particularly the C.P. criticized the Movement on these grounds, but its criticism was a short-sighted polemic rather than a critique of the strategic implications of PASOK's political developments, especially in the context of the Movement's expected entrance into power. Thus, for example the C.P.'s accusations that the Movement was engaging in "compromises with the dominant class" were set in the context of Papandreou's elbow-rubbing with the bourgeoisie on social and other symbolic occasions rather in the context of what its political and economic positions meant for a radical socialist party which was just about to form the government.[41] PASOK did not have a hard time defending itself, as its answer was that its primary concern was the defeat of the "right",[42] which given the dismal economic and social conditions (high inflation and rising unemployment) and attendant social problems in the country, was becoming increasingly unpopular. In this way, PASOK managed not only to confront and downplay the criticisms, but most importantly to lay the ground-work for an artificial polarization between the "right", which had been in power virtually since the war, and the "forces of *allage*," of which it was the most effective representative.

The Organization

It is of course very difficult to distinguish the overall politics of a political party from the politics of its internal organizational structure. As we have seen, PASOK, more than any other political organization, is no exception to this rule. Thus, it is no surprise that the internal life of the Movement reflected, as well as conditioned, the political choices made at the macro level by its leadership. As with the political initiatives of the Movement, during this period PASOK's internal structure did not experience any major changes but rather underwent a further sharpening in the orientation which it had taken after the dramatic events which we examined in the previous chapters. Thus, the Movement's structure remained almost exclusively geared to electoral or rather pre-electoral activity. In fact it remained exceptionally active and constantly reflected the agony of the organization's leadership to expand its electoral support.

While the electoralism of PASOK's organizational structure had been rather tacit up until the 1977 election, its electoral success and its much expected victory forced the Movement to make no pretence about the nature and the role of its organization. To "congregate around PASOK" was now considered "a pressing duty" and the Movement made no mistake in theoretizing its intentions and even justifying the changes in the overall politics of the organization. From all the documents available, it seems to us that an *Exormise* editorial best summarizes and frames the orientation of the Movement's organizational activity. In it they claimed that the form which the "national liberation took to free the country from armed occupation" (i.e. during the Dictatorship)[43] was the correct one. "Now (however) parliamentary and democratic procedures and the popular vote have become its weapons. Populating PASOK's ranks is now the immediate national democratic duty... If it was possible before (for the Movement) to have the flexibility to move with small steps of enlightening and ideological zymosis, today such a flexibility does not exist. New rhythms are becoming imperative...(and) the shape of the new conditions are forcing it to accelerate its steps. If it does not do so, it will be the carrier of heavy national responsibilities... Time waits for no one. The historical moment has arrived. If we missed it we would be responsible to our generation, to the generation of National Resistance and to the generations to come (as well as) to Democracy and the Nation...".[44]

Leaving aside for a moment the implied Burkean notion of society, (since the social project of past, present and future generations is united under PASOK's leadership), this statement crystallizes the organizational orientation as well as the methods for its implementation. First, the Movement indicated that its primary goal was electoral victory and that the strategy for the implementation of this goal required mass population of the organization. It is of course normal for a political party to want to maximize its support, but when we deal with a socialist organization which pursues the expansion of its support in vague and questionable sociological and ideological terms, then a real question is bound to arise as to its actual socialist nature. The case of PASOK, particularly during the period under examination, was a prime example of this rather liberal pattern of expansion.

Many of the Movement's documents indicate that PASOK was gearing towards the election at all costs and its vagueness was becoming surprising even relative to its controversial history. The decisions of its Central Committee fluctuated between calling for making PASOK a "vanguard organization"[45] and appealing to implement the strategy of "National Popular Unity".[46] While none of these goals were defined or elaborated in any detail, the latter stood throughout this period as the Movement's political strategy. However, despite its apparent importance, PASOK did not define it as more than the necessary popular unity for the ousting of the right from power. Thus, the alleged "strategy" remained a slogan open to interpretation of both the organization's internal currents and the population at large.

Furthermore PASOK applied some tactics of mass psychology in order to boost both its credibility and its electoral support. More than ever before during this period the Movement became very chauvinistic about its own organization, political orientation and strength. In many of the party's documents and publications PASOK presented itself as "a unique phenomenon in the history of socialism"[47] with tremendous "historical responsibilities... (which) gave it no right to fail".[48] But even if this organizational chauvinism could be considered necessary for boosting the confidence of the membership just before the election, the Movement exceeded itself when at times its organizational patriotism reached amazing levels. More than once the anticipated boosting of organizational pride became a display of amazing arrogance. For example, an *Exormise* editorial, with a self-congratulatory tone almost reminiscent of Maoist adulation reads "honour and glory to PASOK and to the great leader of the Greek People Andreas Papandreou, who designed the great spiral march of the Popular forces...(in order) to bring them to the threshold of power"![49] This type of rhetoric is rather important point to keep in mind in order to analyse and understand the attitude of arrogance displayed by parts of the Movement's membership (most particularly the infamous "green guards") when the organization formed the government.

In addition, the morale of the Movement's membership was stimulated by advertising the attacks to which various sections of the organization were occasionally subjected by some extreme right wing elements. This victimization of the Movement was provided as living proof that the organization was on the right track in order to implement its historic mission.[50] After all, according to the logic often used in Greek politics and particularly by the left, if your enemy hates you then you must be correct, simply by virtue of this reaction.[51]

Part of the opening up of the organization to the population at large was the different image which the Movement tried to promote through its publications, especially *Exormise*. From as early as the spring of 1978, *Exormise* ceased to be subtitled the "weekly instrument of PASOK" and was transformed into an "independent democratic weekly newspaper". In addition to helping to expand its electoral support, this choice of the Movement *vis-a-vis* its major publications gave further flexibility to the leadership for political manoeuvring which now more than ever before was vital to the implementation of its political plans.[52] The transformation of the Movement's newspaper was not simply limited to its title but it was also reflected in the overall format and themes of the paper. Although it continued to promote PASOK's political line on what it considered political issues, *Exormise* now devoted more pages to a very mainstream analysis of several popular issues (i.e. sports, advice on diet, reports on breakthroughs in technology and medicine etc.).

Initially the transformation of *Exormise* was so dramatic that even the column on "The Life of the Movement" disappeared, and it was impossible to distinguish it from the papers of the liberal press which for some time had turned its support towards PASOK. Soon however (winter 1979) the paper reintroduced the column, although this time with substantial changes in its content. It went from being a column full of the political initiatives and even the theoretical problematic of the organization's various clubs, to a bulletin board announcing the pre-electoral type of rallies in various parts of the country at which the party's top functionaries and more often Papandreou himself were to speak. By the beginning of 1981 when the organization was fully and deeply involved in the electoral campaign (which officially had not yet begun) "The Life of the Movement" had grown disproportionately reflecting the overwhelmingly electoral flavour of the paper. The transformed *Exormise* was in fact the "success" anticipated by its instigators. PASOK's newspaper doubled its circulation,[53] thus further indicating that the Movement was on "the right track" for the upcoming election.

As we saw in the last chapter, in the pre-1977 period, the Movement's organizational activity had an overall electoral orientation especially after PASOK had managed to settle its restless and "dangerous" left-wing currents. During this period the Movement was not tormented by any major internal conflicts or splits (although the 1980 crisis in Salonika might be considered an exception).[54] Since the organization could smell the electoral victory, it was further subsumed this time in a more intense and exclusivist fashion by preparation for the upcoming election and anticipated victory. The mobilization of the Movement's organization and its preparation for the election started almost immediately after the 1977 election at which time the leadership was quick to announce that PASOK was opposed to any form of popular electoral front[55] and that in the "next election (PASOK) will run independently with its own improved programme for big change in the country".[56] It was these sorts of public statements on the part of the Movement's leadership which set the pace for the electoral mobilization of PASOK's membership.

The hyperactivity of the organization continued for over a year-and-a-half the leadership demanded an "emergency alert" from the membership.[57] However, the organizational activity and mobilization of the Movement's membership no longer had any clear link to PASOK's overall political project as it was presented in the "3rd of September" or even as it had been reformulated through the subsequent development of the organization. Again it was lacking any strategy associated with, or derived from the Movement's programme except that of maximizing support for the upcoming election. Thus, the actual functioning of the numerous committees, subcommittees, executive bureaus, at the regional, local and national levels, and the fairly complex, often technocratic tasks assigned to them had one and only one purpose – to further expand the Movement's recruitment and prepare the party apparatus for the election.

In fact the Movement made no pretense as to the nature of this intense activism, claiming that "the basic goals of this continuous mobilization of PASOK will be the shaping of the people, preparation for electoral competition with the right, and the victory of the popular movement"![58] Thus, the membership's activity remained superficial and shallow, as the organizational strength of the Movement was simply measured in numbers and in the degree of efficiency in organizing the leadership's electoral rallies, but never in terms of the political education and awareness of its membership. Even the party's newspaper, although admittedly on the basis of only scattered attempts, did not manage to defuse the rampant organizational electoralism and functioned rather as the informing organizational co-ordinator of the various "activities and problems" of the local clubs.[59] The activities of the local and professional clubs were in fact made up of the organizing of mass rallies at which Papandreou or another top official of the party were to speak and of organizing the logistics for the upcoming election (e.g. postering, canvassing, pulling the vote). Otherwise, the local clubs were primarily, as in the immediate pre-1977 period, drop-in centres for the membership.

During this period however, the Movement managed to further strengthen and consolidate its organizational structure and its recruitment. According to unofficial sources, by 1981, PASOK had displayed a phenomenal increase in its membership which was quoted as 110,000 members[60] (as opposed to the organization's own accounts of 75,000 members in 1980,[61] 65,000 in 1979, and 50,000 just before the 1977 election[62]). Its membership was organized in some 1000 local clubs, 500 sector/occupational clubs and 700 so-called organizational nuclei[63] both in and outside the country.[64] Furthermore, in the 1977-1981 period, the Movement managed to give some life to its regional and national organizational bodies. As we saw in the previous chapter, PASOK's organizational plan was to get organized at three distinct levels: local, regional (Prefectural) and national (Pan-Hellenic). However, until the 1977 election, the plan had been realized only partially since the organization's structure at the prefectural level was rather underdeveloped. As only some of them were formed, most of them were provisional in character and they had not managed to establish any interacting patterns of communication.

During this period however, and in the context of the Movement's organizational hyperactivity, all these committees came into life as numerous Prefectural meetings which were held to elect the respective Prefectural Committees, which in turn started to operate as the communicating channels between the Third Degree or National Bodies of the organization and the Local and Occupational Clubs. Even given their consolidation and development, the prefectural structures were far from becoming the co-ordinating structures for the exchange of the political problematic of the rank and file clubs. They were also far from having any jurisdiction or impact on the formation of the Movement's policies. Thus, given the nature of the Local Clubs and especially taking into account that the jurisdictional boundaries of the Prefectural Committees corresponded perfectly to the country's electoral constituencies, it is only natural to assume that this strengthening of the Movement's activity at the regional level had an electoralist orientation. In fact, the major accomplishment of the Prefectural Committees was the nomination of eight hundred members to make up the Movement's candidacy lists. But even this cannot be considered a major accomplishment since the final decision on the three

The Short March to Power Completed

hundred names needed for PASOK's tickets was left to the Central Committee and to the Executive Bureau which was itself an appointed body.[65]

Needless to say, the combination of electoral anxiety, and (for a socialist party) questionable organizational hyperactivity, gave further strength to the hierarchical structures of the Movement, particularly to the position of Papandreou himself. While Papandreou had in effect become the only source of power within the party structure, and while this development had previously been only tacitly recognized, it was now openly declared. The memberhip was ready for the big day when the winter palace of power was to be stormed electorally. Now, the Movement knew that its electoral success more than ever before depended on its leader and no pretence was made about it. In the political resolutions of the 5th Session of the Central Committee we read "...once more it (the Central Committee) certified that the SINGLE AND STABLE POINT OF REFERENCE in the unsettled consciousness of, and in the desires of the radicals, either among the ranks of people or in PASOK, WAS AND IS A. Papandreou."[66]

This in brief, was the situation in the Movement's organization as it marched with unprecedented arrogance and confidence[67] towards the 1981 election. Although the organizational structure was to work very efficiently in the upcoming electoral battle, it was full of contradictions and controversies which made the Movement's prime electoral slogan – "PASOK in government, the people in power" – sound at the least, ironic, since these contradictions amounted to anything but a guarantee of its implemetation. This organizational controversy along with its programmatic contradictions were soon to be vividly displayed as Papandreou's party moved to power leaving the Greek people once again with the bitter after-taste of lost hope in their struggle for human liberation and a genuine democratic socialism.

The Electoral Campaign and the Pasokean Victory

In an effort to survive the deep changes[68] in its party, the government of New Democracy decided to end its term in power early – the only government in post-war Greece to do so – and announced the election for October 18, 1981. But with a leader (G. Ralles) who appeared weak and often ridiculous as he desperately tried to fill the shoes of Karamanlis, to reconcile the extreme right wingers with the centrist elements in his party, and to deal with an economy in crisis and inflation on an unscrupulous run at an annual rate of 25%, the government did not stand a chance at re-election. Thus, it was becoming clear that the ball in the electoral game was moving into the opposition's range and that the only thing left for the disarrayed right-wing party to do, was to minimize its losses.

The situation within PASOK's ranks was quite different. As we have seen, the Movement was entering the electoral campaign with great confidence and almost certain of the result. It was lead by a charismatic leader who was becoming more and more popular as the party became more and more moderate in its actual political practice. It had also created, after the removal of the radicals, a united, obedient and well oiled organizational structure, whose membership although inexperienced, had high morale and had been well trained. It thus was ready to back up the efforts aimed at the success of the Movement's electoral tickets even in the most remote areas of the country.

In June 1981 the Central Committee of the party met to deal with the "tactics and the strategy" for the election, where it was decided that the Movement was to fight a one front electoral battle – i.e. against the right. Furthermore, its organization was to promote a moderate political climate avoiding extreme confrontations with the other parties.[69] These guidelines, which PASOK's leadership gave to its membership, were a novelty in the context of Greek politics. And since PASOK was one of the major, if not *the* major, participant in the elections, they were to play a important role in the further development of Greek politics, which were entering a new era. PASOK was the first party in Greece which was seriously attempting to oust the right from power without locating its efforts in the context of a "bi-frontal" fight both against the right, and the left.[70] This indicated that old fashioned anti-communism was no longer to be a saleable political commodity. PASOK did not always stick to this tactic during the electoral campaign, and sometimes the public interactions of its functionaries had a flavour reminiscent of the 1960s Center Union "bi-frontal electoral battles".[71] Overall however, the electoral campaign was the most liberal and the "cleanest" in Greece's modern political history.

The Short March to Power Completed

The Movement's political and organizational development, especially in the four years prior to the election, had gained it the title of "responsible opposition party" and even its right-wing adversaries, by 1981, had to admit openly that PASOK was not up to initiating a radical rupture in Greek society since it could not be considered "a revolutionary party".[72] PASOK's moderate tone was set by its continuous appeal to the "underprivileged Greeks" and the need to protect the "lower-middle strata", as well as by the generalities and the minimalist nature of its programme.

The Programme

"PASOK's Governmental Program" did not differ much from its 1977 "Governmental Guidelines". The vagueness of the 1977 electoral programme had been modified, as issues were addressed directly, but the relative concreteness of the new programme was even more detrimental to the Movement's initial radicalism. Contrary to its predecessor, the specificity and the issue oriented programme of 1981 left much less room for radical interpretations and "misunderstandings". The word socialism was not even mentioned as a remote strategic goal.

To say that the programme was concrete, does not mean that the solutions given for the problems were concrete, clear, or free from obvious contradictions. Rather it means that PASOK's 1981 programme focused upon issues which were not related either to each other, or to any strategic goal in any comprehensive fashion.

Still, some of PASOK's original slogans (such as "National Independence", "Popular Sovereignty", "Social Liberation" and Democratic Procedure"), were echoed in 1981, but they were now mixed with a number of other strategic aims which made them unrecognizable. "National Independence, Territorial Integrity, Popular Sovereignty, and Democracy, Autonomous Economic and Social Development, Our Cultural Renaissance, The Revitalization of the Country-Side, The Radical Improvement of the Quality of Life in the City and in the Villages, Social Justice, and finally, Social Liberation"[73] now composed PASOK's strategic goals.

On foreign policy, PASOK's programme promised a "policy

of peace,...realism,...solidarity to the people who struggle for their national independence or...their independent economic development...(and) finally, a policy based on the fact that Greece is a country which simultaneously belongs to Europe, the Balkans and the Mediterranean".[74] On the concrete foreign policy issues, PASOK promised a plan for the gradual removal of the US military bases, it mentioned the possibility of the country's withdrawal from the military wing of NATO (up to that point, it had avoided making the distinction between the political and military wing of the organization) although for the first time this possibility was tied to the simultaneous dissolution of "both cold war alliances – NATO and the Warsaw Pact." And finally, on the hot issue of the country's membership in the EEC, PASOK promised a referendum, but it was quick to point out the possibility of a special agreement.

On the internal front, PASOK stated: "Our fundamental goal is self-sufficient economic and social development, the development of all productive forces in combination with a more just distribution of income and wealth among the various groups in the population and among the regions."[75] Although PASOK addressed the economic issues of the country directly this time, it did not however, manage to overcome its traditional vagueness and confusion, as it clearly tried to include in it something for everyone. The following extract from its programme is a case in point: "The crisis of the capitalist system, on a global scale...the intense competition, the increase at the level of accumulation and the monopolistic structure of many sectors have made the traditional means of economic policy ineffectual. Our choices are based on the application of democratic planning, the restructuring of the public enterprises so that they will become better instruments in support of the private sector, the support of the "lower-middle" businesses, and a nationally advantageous policy for the (attraction) of foreign investment."[76]

PASOK emphasized the need for decentralization, for the transferring of power to the municipal authorities and promised the establishment of collective bodies for the creation and the proper functioning of an "Apparatus of Democratic Planning". But there were no more details given, other than the promise of "decisive participation of the people at all levels and stages...(since) this type of planning, in addition to greater productivity, guarantees the rhythm of a development which is realizable by the people."[77]

The Short March to Power Completed

In addition, PASOK promised a short term economic plan aimed at "controlling inflation, stimulating productive activity, and redistributing income".[78] The promised incomes policy was composed of an "automatic index adjustment" (ATA) on the one hand, and simultaneous price controls on the other. The farmers were promised fair prices, more subsidies, as well as significant increases in their social security benefits (i.e. pension, various types of compensation, etc.). Labour was promised a democratization of the trade union movement with the abolition of anti-labour legislation. Finally, PASOK's 1981 electoral programme reminded the electorate of the Movement's commitment to "the policy of socializations which is a system of social control in some key sectors (of the economy)".[79] However, it was rather quick to point out that "such a policy not only would not undermine private initiative but on the contrary, would stimulate it by promoting the interests of enterprises especially the 'lower-middle' ones".[80]

Furthermore, PASOK's programme made positive reference to a number of other issues which were more immediate and appealing and of which three can be seen as more important than the others. The first was the promise to restructure the chaotic health care system, the second was the promise to democratize the civil service and the third was the promise to solve the ecological problem of the metropolitan area of Athens.[81] Finally, in its attempt to satisfy just about everybody, PASOK's electoral programme made specific reference to the much needed educational reforms, to the arts and cultural activities and finally to legal reforms in the sphere of "equality of the sexes" including the promise of equal pay for work of equal value.[82]

At this point, it should be pointed out that in the context of Greek politics, where actual concrete policies and issues are almost never debated and the content of electoral programmes is rarely juxtaposed, PASOK's programme per se had little to do with the Movement's subsequent success. This was even more the case with the 1981 programme since its language and content (i.e. it still continued to see the solution to the country's problems as managerial ones) was distinctly technocratic. Terms such as "balance of payments", "productivity", or the poorly defined promises of an "apparatus of Democratic Planning" seemed unattainable to the average voter and made a rather inappropriate base for pulling the vote. Thus, both because of this, and thanks to its vagueness, PASOK's programme was hardly ever debated during the election.

If the programme *per se* did not appear to be the focus of the Movement's campaign, its technocratic flavour and especially its technocratic process of creation seemed to be given more consideration, as the origin (professional and not political) of its authors were provided as the proof of its correctness. In the words of Papandreou himself: "I do not think that any other Greek party has gone to such pains...to present a Programme of Governmental Policies as well studied as ours...it took more than a year and a half of work (and)...we have used many people... activists of the Central Committee, members of parliament, the Executive Bureau, myself personally, a significant number of civil servants with long experience in the government, professors from Greek universities, Greek experts who live and work abroad in big international organizations or in universities...(for these reasons), I am personally satisfied and proud of the...result."[83]

In conclusion, it must be noted that PASOK's all encompassing programme was far from containing the totality of popular expectations connected with the Movement's victory. If one is to understand the popular frustration after the first few years of PASOK in government, one must consider all the promises which the Movement had made in a tacit, but much more materially grounded fashion while in opposition. With its active involvement in the mass movement and the oratorical skills of its leader, PASOK's political discourse since 1974 had created expectations about the dreamed *Allage* that even PASOK on the eve of the election had not anticipated.

Allage

Ironically, what really served as the foundation of the much advertised moderate political climate was PASOK's main slogan while in opposition, that of *allage*. This slogan, which was not completely unknown in the post-war political history of the country,[84] became the common political denominator of the entire spectrum of the opposition from the moderate parties of the Center to the C.P. However, through its political practice, PASOK made the Movement the main if not the exclusive advocate of "all the tendencies which want *allage*".[85]

"The time has come for the great *Allage* in our country", PASOK declared in the opening line of its "Declaration of Governmental Policy" or as it was usually called "The Contract with the People", but this slogan was anything but concrete and further was not even remotely associated with the project of socialism. In its vagueness and simplicity *allage* for the people of Greece in 1981 had become the crystallization of a diffused, instinctual expression of their desire to oust the right from power. In other words *allage* had a predominantly negative content. It was a content which was very capable of attracting all those who were dissatisfied with the present state of affairs (since by definition it was "progressive"), while at the same time it was neither overly alienating nor overly exclusivist. It could be given almost any content which the electorate wished to give it.

However, there was no doubt that the popular feeling for change was very strong and that PASOK appeared to be the only party capable of delivering it. PASOK, though, in spite of its previous (potentially) socialist premises, did not manage or rather did not attempt to give this popular desire for change a radical content. Of course according to the Movement's programme: "The conquering of national independence, of popular sovereignty and social liberation, and the socialist transition, compose the vision of the great *Allage*" but in the context of the overall programme and electoral discourse of the party these had anything but socialist connotations. In fact "socialism" appeared to have been put on hold since it was now regarded as a remote goal and a "vision" which was not on the present-day agenda. Instead, the socialist ideal was overpowered and eventually covered up by the general, vague term *allage*. As long as this term remained undefined, removed from, and unconnected to the concrete changes which the transformation to socialism entails, it was bound to remain simply a negative and overly malleable slogan. In addition, PASOK's social references to "underprivileged" were by definition inhibiting the development and the articulation of the concrete socio-political discourse necessary for the implementation of a socialist vision. A political message, a political slogan to convey genuine socialist politics, even as a remote long term possibility, has to be articulated and organized at both the socio-economic and the political levels with processes which are objectively conducive (i.e. open and democratic) to a radical restructuring of the society (socialism).

However, the condition and the overall non-democratic nature of the Greek parties of the opposition and their non-class fashion of conducting politics, made the radical expression of the otherwise powerful *allage* very unlikely. There was a great deal of loud talk about the need for displacing the right from government but only whispers, at best, about socialism. For a case in point, it is not by accident that from PASOK's admittedly extremist slogan of "socialism on the 18th of November" in 1974, we reached the 1981 electoral rally of the party at which, in spite of its anti-rightist tone and the radicalism of some long standing slogans (i.e. "EEC and NATO the same syndicate" etc.), the word socialism was not even mentioned once.[86] Thus, and in order to put the unprecedented 48% of the popular vote that the Movement was awarded by the electorate into perspective, it is only natural to conclude that the social mosaic which supported PASOK had a liberal (centrist) orientation, similar to the origin of the majority of its membership and that they were driven primarily by their desire for *allage* and not by a vision of socialism.

The Left

Although PASOK was undoubtedly the major contributor to the moderate definition of *allage* and to the overall "moderate climate" during the pre-election period, it was by no means alone in this. Obviously no one would expect the right or the other parties of the liberal Center, or even those with a social-democratic flavour to resist this watershed of moderation and the "non-definition" of *allage*. However, one would naturally have expected the communist left not only to seriously criticize PASOK's new political orientation, but also to attempt a different definition of politics, or at least to object to the exclusively moderate parliamentary interpretation of *allage*. But the communist parties, in particular the traditional C.P. which was stronger and was not struggling for its survival like its counterpart "of the Interior", failed both to effectively criticize PASOK and to develop a certain discourse/practice to counter the way in which Papandreou's party was moulding Greek politics.

The Short March to Power Completed

Early on, the C.P. had foreseen that the growing popularity of PASOK would become the snowball which was to push the Movement into power. However, as it became more and more clear that this popularity was accompanied by a growing compromise of its radicalism, the C.P. did not manage to develop a coherent criticism of Papandreou's party. Sharing similar views with PASOK on more substantive issues, (basically the conception of the dependent nature of the Greek social formation), the C.P. in its criticisms found it difficult to go beyond warnings of the "dangerous" and "divisive" implications of the "slogans of some democratic forces to form a majority government"[87] or the questioning of PASOK's leadership's social contracts with known members of the Greek bourgeoisie.[88] But even when the C.P. got around to putting forward a more substantive and theoretical criticism of PASOK, its criticism was not any more effective. In its theoretical encounters with the Movement, the C.P. did not focus its criticisms on what PASOK actually was or actually did, but rather on what PASOK might have been and/or what Papandreou's party did not talk about. Thus, in these analyses or rather sophisticated polemics, "the programme of PASOK stopped where it was supposed to have started" and one of the prime vices cited was its failure to externalize its already internally adopted principle of democratic centralism.[89] If criticism is not conducted at the level and within the parameters of the subject matter, then the attempted criticism runs the danger of becoming idealistic or materially unsound, and thus it is bound to have no effect, since it usually "transcends" itself into an isolating monologue.

But if there is a single area in which the C.P. can most rightfully be blamed for being conducive to the above described developments in the political climate, this is its actual political practice. In its effort to avoid the crushing effects of the growing polarization between New Democracy and PASOK, and the devastating mass psychology of the "lost vote" the C.P. concentrated its political activity exclusively on its own electoral success. In fact in the 1977-1981 period, the "party" became literally obsessed with the upcoming elections. Making the overly ambitious goal of 17% its guiding flag and the general and no more concrete call for "real *allage*" its prime political argument, it entered the electoral campaign like any other main-stream political party. In this way the C.P. managed to foresake its natural capacity, (since it was/is the main courier of the communist tradition in the country), to radically redefine the existing political discourse. Thus, it was no surprise that the "theodicy" of the electorate avoided the uncertainty of "real change" and instead opted for the certainty of the winner (PASOK) giving to the C.P. a mere 10.94%.

For its own part, the C.P. of the Interior did not fall short in its contribution to this moderate, narrowly defined electoralist and "classless" definition of politics. Having broken the "Alliance" – the coalition it had initiated in the previous election – the party decided to try its luck at the polls on its own. Given its miniscule size, it tried to break the polarization of the major electoral competitors and survive the parliamentary game by acting as the dam. It claimed that "it was impossible (for PASOK) to gain the absolute majority" and therefore a vote for them was at least as effective.[90] This attitude however, and the practice which resulted from it, along with the avoidance of any real challenge or even rhetorical criticism of PASOK's practice[91] did nothing but further contribute to the "moderate climate" as it was basically defined by the leading opposition party. Thus, it was no surprise that the "Interior's" failure to differentiate itself from the main-stream politics of the opposition cost it its parliamentary survival. Winning only a pitiful 1.35% at the polls, the "small C.P." found itself outside the parliament.

The Campaign

If political moderation crystallized by the simplistic slogan of *allage* (which was the common denominator of almost all the parties[92] participating in the election), was of great significance for the future development of Greek politics, it did not stand alone. The style and the method of conducting the electoral campaign was the other significant characteristic of the 1981 elections which was to develop into a permanent trait of electoral competitions in the country. Here the common denominator of the parties' electoral tactics was the "commodification of the electoral antagonism".[93] When politics are neither connected to their social base nor to concrete social projects in any structured, direct and conscious way, they are then bound to be confined to public relations, which in turn become subject to marketing techniques.

The Short March to Power Completed

When, under the supervision of big advertising companies, the parties were competing to see which would fill more available space with colourful posters, or which would put together more impressive rallies, or whose loud speakers were louder or even which party would manage to initiate (usually from the balconies) smarter slogans to attract the attention of the masses, they were radically redefining not only the electoral game but also Greek politics in general. The people were being treated as passive spectators of electoral fireworks, and electoral slogans were replacing political arguments. Politics as a meaningful social activity was debased into an instinctual and sensationalized interaction. Thus, the electoral/political arena in which the crystallization of social conflicts takes place was transformed into a football stadium from which the fans returned home for a beer after a two hour emotional release. Of course this criticism is not to imply that politics, as in fact every human activity, even in ideal situations, does not, or should not, entail a certain amount of emotionalism. But things had gone too far. Even if this was not obvious at the time, since the pattern had just been inaugurated, it did not take long (i.e. the election for the European parliament 1984) before even its most enthusiastic instigators became fed up with its most crude expressions.[94]

Finally, the sensationalization of Greek politics had a rather decisive impact on the party system. Since the marketing method of conducting politics requires tremendous financial resources and in a sense, blackmails the electorate, it promotes the bigger political parties while it squeezes out the smaller ones. Thus, it further distorts the articulation of political representation. This has been particularly important because Greek politics had traditionally operated in the context of a multi-party system. But furthermore, given the country's social configuration, the absence of democratically run "multi-tendency" political parties and the disadvantageous consequence of the electoral law on small parties have had a devastating impact on democracy itself.[95]

The Result

The unifying effects of the moderate interpretation of *allage* gave PASOK all the legitimacy it needed for filling up its lists with politicians who just recently had fled the hopeless parties of the center and had become backroom supporters of the Movement's *allage*. But while the surprising mass exodus of the politicians of the center had been taking place for some time, on the eve of the election it took scandalous dimensions as G. Mavros triumphantly gained a prominent position on the Movement's lists.[96] Mavros was the leader of the remnants of the old Center Union in the post *metapolitefse* era, and had served as Deputy Prime Minister and Minister of External Affairs in the government of "National Unity" that Papandreou himself had called a traitorous configuration. At the same time, Papandreou[97] also had to take advantage of the naivete and why not opportunism of the left and counterbalance it with the simultaneous acceptance of a prominent left wing cult figure M. Glezos[98] onto the Movement's list.

When the first results of the election were announced on the morning of 19 November 1981,[99] no one was really surprised. With a commanding 48.07% of the popular vote, PASOK had gained 172 seats in the 300 member parliament. It had increased its electoral support since the last election four years earlier by a remarkable 22.73%. New Democracy had shrunk by 5.96% to 35.88% but in effect, its losses were heavier if one considers that its victorious performance in 1977 with 41.84% had occured when the extreme right wing had run independently and won a shocking 6.82% of the vote. In the 1981 election however, many of the ultra-right wing candidates had been quietly included in New Democracy's lists, allowing in this way, a mere 1.69% of the vote to fall to the extreme right. The Communist Party fell far short of its ambitious 17% goal, but increased it electoral support by 1.58%, gaining 10.94% and 13 seats in the parliament. The C.P. of the interior ran independently to win a mere 1.35% and to find itself outside of the parliament. However, the really big loser of the election was the center which was in effect obliterated electorally with only 0.40%, a decline of 11.55% from the previous election.

Given that the traditional left had in effect remained stagnant, it would be reasonable to assume that PASOK's support had come from the right and primarily from the Center's 11.55%. There is no doubt that PASOK's new moderate image, which was accompanied by the inclusion in its lists of many of the turncoat politicians of the center, had paid off. At the same time however, the inclusion of the ultra-right in New Democ-

The Short March to Power Completed

racy, combined with Karamanlis's departure from its leadership seems to have alienated a good part of the liberal section of the party, which apparently decided to vote for PASOK and its charismatic leader.

What made PASOK's victory even more impressive was the distribution of its vote. Its 1981 electoral support and victory was national in scope. It was far from concentrated in certain areas of the country. In fact PASOK only lost by a very slim margin in nine out of the country's fifty-six constituencies, breaking down the long standing bastions of right-wing support. The establishment and the expansion of the Movement's Local Clubs especially in the country-side can be seen as a major contributor to this success.

Furthermore, the social origin of PASOK's supporters appears to have been quite diverse, maintaining a strong presence, if not the majority, in all the social categories used at least in one survey (see Table I below). Although it used anything but Marxist categories for its base, this survey with the breakdown of the social support of the Greek political parties, has been used to "prove" the working class nature of PASOK.[100] It is true that PASOK had won an absolute majority among the "workers" and the "wage earners" (unfortunately the survey does not indicate the difference between these two categories), but class electoral support is not enough to characterize a party as working class or otherwise. Such an assumption would amount to a mechanistic "sociologism". In the case of PASOK, even if we put aside the tremendous support which the Movement enjoyed among senior executives and managers (62%) as well as among professionals (40%), or for every party for that matter, and even if we assume that such worker and earner electoral support is a good indication of working class participation, class participation alone is not sufficient to define the nature of a political party.

Unless we considered the political discourse in combination with the class participation, it is rather questionable if we can make a reasonably safe claim as to the party's nature. In this case, PASOK's political discourse (i.e. its appeal to the underprivileged, its nationalism, its centralized and personalized structure) was hardly a working class one. Thus, any claim for the working class nature of PASOK is bound to be far-fetched.

The Short March to Power Completed

Table I: Percent Vote by Occupational Category

Occupation	PASOK	N.D.	Other	Total
Business Owners with employees	13	56	31	100
Business Owners no employees	26	44	30	100
Self-Employed Professionals (Doctors, Lawyers, etc)	40	23	37	100
Senior Executives (Managers, Professors)	62	31	7	100
Self-Employed Non Professionals (Technicians, Shopkeepers)	48	29	23	100
Wage Earners	53	27	20	100
Workers	57	17	26	100
Farmers	47	38	15	100
Pensioners	32	49	19	100

Source: Survey Conducted by the Nielson Co., quoted in *Marxistiko Deltio* (Marxist Bulletin) Vol. 13-14, 1982.

Conclusion

Prior to the election, PASOK had confidently called the anticipated victory its "rendez-vous with history". But even on the night of the election, the cheers for the end of the almost 40 year right wing reign could not hide the fact that the "rendez-vous" was not going to be a happy one. PASOK inherited a rather dismal economic and political situation from its predecessors. Inflation was running at more than 25% per annum and real wages had declined by 5.5% from 1979-1981; although not officially recorded, unemployment was becoming visible among young people; there was a lack of investment (especially in the primary sector) when for the first time in the post-war history of the country there was negative growth in the GNP; there was a growing trade deficit; invisible foreign exchange receipts registered a drastic drop (due to the crisis in sea transport and tourism); and there was a deficit of $2.3 million in the balance of trade.[101] In addition, PASOK inherited a huge, inefficient, inflexible civil service, and a number of long term problems (the education system, the justice system, and the pollution crisis in Athens).

It had been predicted that PASOK's major adversary was

not going to be so much the dismal situation just detailed, but rather that once in power it was going to inherit the harvest of its own programmatic political seeds, which it had planted while in opposition.[102] PASOK had been promising too many things, not rarely contradictory, to almost every possible constituency and it was just about to face the consequences. How, for example could the new government overcome the "one-sided austerity" performed by the right and for that matter any austerity when they had promised that all the plausible demands, with the exception of the ones of the "international and national monopolies", would be fulfilled? How could the stimulation of the investment climate be achieved through public spending and other incentives without raising taxes or prices? How was it possible to cancel the country's membership in the EEC and then force the community to sign a "special agreement" which would give the country the right to do anything it wanted with the Community's support? How could the government respond positively to numerous sectoral and often sectarian demands of various social groups (from lawyers to taxi drivers, and from small businessmen to environmentalists) and at the same time introduce rational economic and social planning? How could the plans for "democratic and decentralized planning" have any chance when the new governmental party, and all other parties for that matter, were profoundly undemocratic and centralized? These were some of the questions deriving from the endless list of PASOK's contradictions that the new government would be called upon to face. But this is something with which we will deal in the next chapter.

Notes

1. Title of an editorial in the first *Exormise* published after the election (25 November 1977). In the same editorial it is revealed that in the Second Session of the Central Committee, where electoral tactics were discussed, P. Avgerinos, a member of the Executive Bureau, had stated that "the (electoral) goal of PASOK is to double its electoral power and to elect at least one M.P. in each constituency". PASOK had in fact done nothing less than that.

2. It is exactly for this reason that we will avoid referring to all of these withdrawals and expulsions of PASOK's membership. There was constant discontent among the Movement's membership which rarely

reached the level of extensive membership withdrawals. The most significant incident of this kind, as we will see, were the dismissals which occurred in the regional organization of Salonnika in the summer of 1980, but even this case, which included top local activists and one member of the Movement's Central Committee, failed to have a significant impact, and was soon forgotten.

3. Second Session of the Central Committee.

4. Editorial on "Seven Former M.P.s who Joined the Movement" *Exormise*. 11 February 1979.

5. Ibid.

6. See A. Papandreou, "Proposal to the Central Committee" *M.I.B.* July-August 1978, in which the original radical opposition to the "allies" is maintained fully. Also A. Christodoulides (member of the Executive Bureau) "Na Ekfraste Sosta e Ethinike Omopsechea" ("To Express Rightfully the National Unanimous Will") *Exormise* 7 October 1979.

7. Editorial in *Exormise* 4 June 1978.

8. *Exormise* 16 April 1978.

9. See for example: J. Papaspyrou, "To Emporiko mas Naftiko Synthlivete apo ten E.O.K." ("Our Commercial Marine is crushed by the EEC") *Exormise* 23 and 30 June 1979; also in an editorial in *Exormise* 31 May 1981, we read that "...the community regulates our foreign policy...(and) it is leading us to a confrontation with the Arabs...(since it) requires a full relationship with Israel". When one considers the close economic ties of the relatively weak Greek industry with the Arab countries, the point is well taken.

10. The Movement rejected the proposal of the Socialist International to join, on the basis that the S.I. had accepted Etzevit's party (the Socialist Labour Party of Turkey) as a member. See: A. Christodoulides (member of the Executive Bureau) "Ten Politike ton E.P.A. Ekfraze e Sosialdemocratia" ("Social Democracy Expresses US Politics") *Exormise*. November 12, 1978.

11. See: A. Christodoulides "Me ta Opla tes Economikes kai Technologikes Voitheas sto Trito Kosmo Stochos tes Sosialdemocratias" ("With the Weapons of Economic and Technological Aid Social Democracy Aims against the 3rd World") *Exormise* 5 August 1979; also M. Charalambides (member of the Central Committee) "E Chores tes Mesogeou o Stochos tes Sosialdemocratias" ("Social Democracy Aims at the Countries of the Mediterranean) *Exormise*. 11 and 18 February 1978. It is worth noting that both the authors of these articles and other prominent members of the Movement were pushed over to the fringes of the organization in 1983.

12. See for example the editorial article "The Black Bible" in *Exormise* 18 February 1979, in which a list of some prominent members of the Greek bourgeoisie who declared very low taxable incomes was

published.

13. *Exormise* Editorial comments 10 May 1981. The comment referred to the President of the National Association of Agricultural Co-ops.

14. *Exormise*, headlines and main editorial article, 12 September 1980.

15. A. Papandreou, interview in *Exormise*, 30 August 1981.

16. See the presentation of J. K. Galbraith's article "Plouralistikos Sosialismos o Monos Dromos Soterias" ("Pluralistic Socialism the only Road to Salvation") in *Exormise*. 26 November 1978. For other examples of this kind see the lists of proposed readings for the membership by the Center of Research and Enlightenment (KE.ME.DIA) published repeatedly in *M.I.B.* and *Exormise*.

17. See for example the glorifying report on "Enas Chronos Apophasistikes kai Skleres Maches tou PASOK ste Voule" ("A Year of Committed and Tough Battles of PASOK in Parliament" *Exormise*. 31 December 1978. Also "Oi Vouleftai tou PASOK ste Mache yia ta Provlimata tou Laou" ("PASOK's M.P.s in the battle for the problems of the people") first page editorial in *Exormise*. 26 November 1978.

18. See article by the previously ultra leftist M. Nikolinakos "E Praxe prepe na Syntheetai me te Theoritike Vase" ("Praxis should be Connected to the Theoretical Base"), a three part article in *Exormise*. 24 February – 10 March 1978.

19. "The 5th Session of the Central Committee of PASOK: Towards a Mass and Vanguard Organization" Athens, PASOK Publications, February 1979, 2nd edition 1982. p. 38. (emphasis in the original).

20. Under the peculiar modified proportional representation of the Greek electoral law, at the time, for a party to be allowed to participate in the second distribution of votes and therefore possibly to gain parliamentary seats, it must win at least 17% of the national vote. Well before the 1981 election, the C.P. put forward the slogan "K.K.E. Allage – B' Katanome" (C.P. Change – 2nd Distribution of Votes) which literally dominated the entire existence of the party, reducing every aspect of the party's politics to it.

21. A. Papandreou "Speech in the Parliament on the 1978 Budget" in *M.I.B.* May 1978.

22. D. Zachariades-Souras "E Pathitike Exartise apo Diafores Peges: E Kapitalistike Anaptixe tes Ellinikes Viomechanias" ("Passive Dependency from Various Sources: The Capitalist 'Development' of Greek Industry" in *Exormise*. 22 November 1980. For a problematic similar in nature see M. Beteniotes "To travma tes nafilias to Pleronoun oi Naftike mas" ("The Wound of Shipping is Paid by our Marine Employees") in *Exormise*. 3 February 1980, (the title has nothing to do with the content as the criticisms of the shipowners are no longer anticapitalist or even nationalist in nature but rather highly technical and

focus on tax laws). Also see articles in *Exormise*. October 1980; and an investigative report on the activity of foreign banks in Greece in *Exormise*. 16 December 1979.

23. For the Movement's rationalization of the vote see *Exormise*. 4 October 1979.

24. Interview with D. Tzouvanos (then a member of the Central Committee) op. cit.

25. See *Metavase sto Sosialismo* (Transition to Socialism). Athens, Aletri, 1981; and *Ellada kai Sosialismos* (Greece and Socialism), Athens, Kastaniotes publications, 1982.

26. This is the main thrust of the debate among various prominent members of the Movement published in the major publication of the independent left in the country. *Anti*. No. 122, 123, 124. For an overall analysis of this problem see: M. Papajannakis "To PASOK apo ten Antipolitefse pros ten Kyvernise" ("PASOK from the Opposition to the Government") in P. Papasarantopoulos ed. *PASOK kai Exousia* (PASOK and Power) Athens: Paratiretes, 1980. pp. 431-445.

27. See for example the article by A. Papandreou "To Ethniko Adiexodo kai e Antimetopise tou: Exi Protases yia To Xeperasma tes Crisis" ("The National Dead-End and its Confrontation: Six Propositions to Overcome the Crisis") in *Exormise*. 28 October 1980.

28. A. Papandreou's proposal to the 8th session of the Central Commmittee in *Mprosta stes Ekloges kai ten Nike* (Confronting the Election and the Victory). The 8th Session of the Central Committee of PASOK, Athens. PASOK publications/KE.ME.DIA., 1981 and *Exormise*. 30 August 1981.

29. Ibid.

30. Theoretical article written and distributed within the organization (21 June 1978) by KE.ME.DIA. (Center of Research and Enlightenment) "Ta Micro-mesea Stromata e Rachokokalia tes Allages" ("The Lower-Middle Strata the Back Bone of the Change") published in *M.I.B.* July-August 1978.

31. Ibid.

32. See for example *Exormise*, Main article 2 March 1980; and A. Papandreou's interview in *Exormise*, which, in more than one way made it clear that "PASOK will guarantee a smooth change in power". (30 August 1981.)

33. *Exormise*. 23 August 1981.

34. This was and is true for both the right and the left of the political spectrum. But whereas right-wing leaderships tend to give emphasis to good public relations with their Western European counterparts, the left traditionally put forward foreign models of development as the living legitimizing proof of the correctness of their political strategy.

35. This was the phrase which the editors of *Exormise* had chosen in announcing Papandreou's trip to Sweden and Norway in the spring

of 1978. *Exormise*. 14 May 1978.

36. See for example the reports in *Exormise* on Papandreou's trip to the Soviet Union (5 October 1981); on his trip to Bagdad (9 March 1980); on his trip to Paris (22 March 1981); on his trip to Bonn (5 April 1981); as well as the report on the Labour Party's visiting delegation (12 April 1981).

37. In August 1981, Soares, Craxi, Gonzales and Zospen (of the PSF), responding to PASOK's invitation, met with Papandreou in Rhodes to discuss issues concerning "the socialists of Southern Europe".

38. *Exormise*. 30 August 1981. Also in Papanderou's proposal to the 8th Session of the Central Committee in *Mprosta stes Ekloges kai ten Nike* (Confronting the Election and the Victory) The 8th Session of the Central Committee of PASOK, Athens, PASOK Publications/KE.ME.DIA., 1981. There were many positive things to be said on the record of the French socialist government, at least during its first year in power (see: G. Ross and J. Jenson "Crisis and France's 'Third Way'" in *Studies in Political Economy* No. 11. Summer 1983; and J. S. Ambler(ed.), *The French Socialist Experiment*. Philadelphia: Institute for the Study of Human Issues, 1985). However, the uncritical advertisement of the French socialists, given the differences with the French case as well as the up to that point polemical attitude of PASOK towards them, can be seen only in the context of the Movement's attempts to further legitimize its own image within the country.

39. Interview with D. Tzouvanos. op. cit.

40. PASOK always voted with the government in support of the ever increasing military budget, as the "armed forces" and only they were considered to be "the backbone of National Defence". (A. Papandreou, speech in the Parliament *Exormise*. 7 October 1979). Similarly, up to that point Papandreou never failed to compliment the army and its role especially during the celebration of the national days by overplaying their role in the glorious victories of the nation, provocatively ignoring the peoples' role in them. See *Exormise*. 28 October 1979, and dailies.

41. See editorial by Ch. Florakis, secretary general of the C.P., in *Rizospastis*. 12 March 1978.

42. See for example, editorial note in *Exormise*. 4 November 1979, by the title "O Monos Stochos tou PASOK e Dexia" ("The Only Target of PASOK is the Right").

43. This is a classic example of the case in which the model does not fit the historical facts and therefore the historical facts have to be changed to fit the model. As we have said PASOK's radicalism was so profoundly attached to the third world dependency type of model that it perceived the dictatorship to be a period in which the country was under foreign occupation!

44. *Exormise* first page editorial 14 January 1979.
45. Decision of the 5th Session of the Central Committee on the organizational orientation of the organization. *M.I.B.* March 1979. Also see *Exormise*. 4 March 1979 Papandreou's article which was used as a preface to the publication of the documents of the session.
46. 2nd Session of the Central Committee 25-26 February 1978, resolutions of which were published in *M.I.B.* March-April 1978.
47. Statement of A. Papandreou, quoted in *Exormise* 3 March 1980.
48. Editorial *Exormise*. 6 April 1980.
49. First page editorial *Exormise*. 18 January 1980.
50. There were many incidents of extreme right wing attacks on the offices of progressive groups, especially from 1979 onwards. PASOK not rarely became the object of such attacks and subsequently did not miss the opportunity to react to the incidents, less in an effort to protest to the undermining of democratic rights, than to advertise the organization's efficiency. See the prominent publication of these events in the pages of *Exormise* in 1979. Also see the arrogance of the report on the *Exormise* trial when the newspaper had to go to court for violating the very strict regulations of publishing. op. cit. first page 21 October 1979.
51. The left and particularly the C.P. practiced this logic more than any other part of the spectrum and did not miss the opportunity to gain points in the sphere of public relations either by taking pride in the historical suffering of leftists in the country or by advertising the current attacks.
52. This was/is allegedly the reason that the leadership has discouraged the establishment of the organization's own daily newspaper.
53. It is rather difficult to be more precise than this so far as the actual circulation of *Exormise* is concerned since in its effort to show its strength, the organization, was anything but consistent in reporting the paper's popularity. On 3 March 1978, *Exormise* reported 40,000 issues as its circulation, on 10 March 1978 it claimed 45,000, on 16 April 16 1978, when the transformation of the paper had taken place, 100,000, while on 4 June 1978 it claimed a circulation of 70,000.
54. As we have seen, there was a strong rebellious tendency in the Salonika section of the Movement. Through the appointment of faithful Prefectural Committees and its attempts to control and diffuse the youth (radical) presence (no more than 7 youth members were allowed in the local clubs), the Movement's leadership tried unsuccessfully to control the situation. Finally, the most active members of the region gathered around L. Tziolas, (a city hall alderman who had managed to get elected to the organization's Central Committee) and pursued a development somewhat autonomous from the central bodies of the Movement, and most importantly backed Tziolas' stands in the Central Committee's sessions. Thus, in the 6th Session of the Central Commit-

The Short March to Power Completed

tee (op. cit.), Tziolas, expressing the reservations of his comrades regarding the new direction of the organization, abstained from voting on the political resolutions of the session. The situation became really hot when in the 7th Session of the Central Committee (13 July 1980), where the "uniformed conception" of organizing the Movement's practice was pressed upon the organization (see the documents of the "7th session of the Central Committees of PASOK" in *PASOK in the Government the People in Power*, op. cit.) and Tziolas voted against it. This friction reached a dead end during the preparations for the Movement's youth festival when the leadership, using the excuse of the "alienation" of the nature of the festival, dismissed Tziolas and two other activists. However, the success of the festival of the "dissidents" meant to the Movement's patriarchs that more drastic measures had to be taken. At the end of the internal fights (3 December 1980) which followed and which not rarely took familiar ugly expressions, 99 activists had been expelled from the Movement and hundreds others had left it in disappointment.

55. A. Papandreou quoted in the first page of *Exormise*. 21 May 1978.

56. *Exormise*. 18 February 1979.

57. See main title in *Exormise*. (6 April 1980) as the inauguration of the electoral campaign was announced by Papandreou and the Executive Bureau.

58. First page editorial in *Exormise*. 4 May 1980.

59. On 9 September 1979 *Exormise* announced a weekly report on the "political strength of PASOK", which in effect amounted to a reporting of the "activities, the presentation of problems and the membership's reports". These reports stopped three weeks later without explanation, however it became obvious that the organization was so much subsumed by the terminal condition of the electoral campaign that there was no material of the kind originally prescribed coming from the local and occupational clubs of the Movement.

60. This is an approximate and unofficial figure of the Movement's membership quoted in the daily *To Vema*. 13 May 1984. In spite of our efforts to get official figures of PASOK's membership and its social composition, the Organizational Bureau of the Movement claimed that it had no such data.

61. Quoted in B. Kohler. *Political Forces in Spain, Portugal, and Greece*. London: Butterworth Scientific, 1982. p. 30.

62. This was the Movement's membership as it was quoted in A. Papandreou's preface to the publication of the documents of the 5th Session of the Central Committee. Athens, PASOK Publications, 1982 (2nd edition). Also published in *Exormise*. 4 March 1979.

63. The Movement's organizational nuclei could not elevate themselves to Local clubs unless they had more than 25 members (in the

urban centres) or 10 members (in the countryside). *PASOK Katastatiko* (PASOK Constitution). Athens: PASOK Publications, 1976.

64. We should not forget that even until today, the Movement has maintained and in some cases expanded its presence abroad (especially in North America and Australia). This is a structure which is both its heritage from the "Junta Years" and a product of large Greek immigrant communities through out the world.

65. *Exormise.* 4 May 1980 (main title).

66. 5th Session of the Central Committee op. cit. p. 192. (Emphasis in the original).

67. Especially after 1979, *Exormise* is full of deterministic and prophetic statements regarding the outcome of the election. Statements of the type "PASOK on the threshold of power" (Papandreou op. cit. 2 September 1979) or "No one can stop the march of people to power" were common in the Movement's jargon.

68. In the Spring of 1980, Karamanlis, the acclaimed "father of modern Greek democracy" and the founder of New Democracy, became disillusioned with his party's contradictions and the accumulated problems of the country. Foreseeing the coming of PASOK's victory, he moved from the prime-ministership to the presidency. From this position, given the powers assigned to the position by the constitution, he hoped to keep on top of things, at least for another five year term and in this way to counter-balance the expected changes of Papandreou's *allage*. His departure from the party however, brought into the open the differences and the contradictions in New Democracy – namely the antitheses between the modernists with a Keynsian orientation and the "old guard" composed of extreme right wingers – that with his charisma he had managed to keep in line since the *metapolitefse*. Thus, the dominant right wing party had to regroup and solve its internal contradictions and this was something that, at least temporarily, New Democracy managed to do remarkably smoothly for a major Greek political party. New Democracy became the first, and until today the only major political party in the country which changed leadership in a democratic way (although some might question the democratic nature of the procedures) without major disturbances to its unity.

69. "Mprosta stes Ekloges kai ten Nike" ("Confronting the Election and the Victory"). The Meeting of the 8th Session of the Central Committee of PASOK Athens, KE.ME.DIA./PASOK Publications, 1981. See also: K. Laliotes, "E Eklogike Tactike tou PASOK" ("The Electoral Tactic of PASOK") in *Exormise*. 30 August 1981.

70. To prove the point all we have to do is to remind ourselves of the fierce anti-communism which accompanied G. Papandreou's efforts to defeat the right in the mid-1960s.

71. See for example A. Strates, "To Programma tou PASOK Apandise sten Polemike tes Dexias kai tes Paradosiakes Aristeras"

(PASOK's Program, an Answer to both the Right and the Traditional Left). *Exormise*. 2 August 1981; also see editorial note in ibid., where the criticisms of *Vradine*, a prominent right wing newspaper and *Rizospastes*, the C.P.'s daily are discussed as if they were stemming from the same point, since they were both engaging in "an anti-PASOK campaign".

72. From *Kathimerine*, a prominent right wing daily, quoted in D. Androulakis, "Ta Kommata kai oi Stratigikes ton Taxeon" ("Parties and the Strategies of Classes") in *Communistike Epitheorise* (Communist Review). March 1978.

73. *To Kyvernetiko Programma tou PASOK* (The Governmental Program of PASOK). Athens, 1981.

74. Ibid.
75. Ibid.
76. Ibid.
77. Ibid.
78. Ibid.
79. Ibid.
80. Ibid.
81. Ibid.
82. Ibid.

83. Papandreou's proposal to the 8th Session of the Central Committee of the Movement. op. cit.

84. The slogan of *allage* had been used in the immediate post civil war period by the National Progressive Center Union (Ethnike Proodeftike Enose Kentrou-E.P.E.K.) under the leadership of Plastiras in an effort to rally those previously recruited by the resistance and to displace the right from power. See: J. Meynaud, *Politikes Dynames Sten Ellada* (Political Forces in Greece). op. cit. pp. 85-94.

85. A. Papandreou, Interview in *Ta Nea*, February 28, 1978.

86. A. Papandreou, speech in the final electoral rally for the 1981 election, 13 October 1981. Personal tape recording.

87. See for example editorial commentary in *Rizospastes*, 9 September 1980, where a criticism of PASOK is attempted but the latter is not even mentioned by name.

88. Ibid.

89. D. Androulakis, op. cit.

90. See for example the editorial note in the party's daily *Avge*. 9 September 1980.

91. In fact the truce between these two parties in this period became monumental. Since PASOK visualized the C.P. of the interior as the only party capable of limiting the strength of its major challenger (i.e. the C.P.) while the latter was terminally preoccupied with the unrealistic application of Eurocommunist models in the country, the two parties were on rather friendly terms. In fact, even on the eve of the elec-

tion, PASOK went so far as to state that "So far as its (the C.P. of the Interior's) presence in parliament is concerned, not only do we not oppose it but we also hope for it". Editorial note in *Exormise*. 2 August 1981.

92. Of course there were a number of small parties, especially those with so-called ultra leftist views, which did not fall into this category. However, their marginal status and activity made their impact by definition negligible.

93. A. Manesis, "E Proeklogike Diadikasia: Programmata, Tactikes, Yfos" ("The Pre-electoral Procedure: Programs, Tactics, Style") in N.P. Diamantouros, P.M. Kitromilides, G.T. Mavrogordatos (eds.) *Oi Ekloges tou 1981* (The 1981 Election), Athens, Hestia, 1984. p. 21.

94. The commodification of Greek politics had reached such shocking plateaus that just two and half years later (election 1984) even Papandreou himself had to publicly cry (although of course after the election) "never again".

95. In the 1985 election for example six parties had to forget or in effect hide their differences and merge with the three major parties in order to achieve mere survival.

96. Mavros headed the list of names of twelve candidates who, according to Greek electoral law did not have to run in any particular constituency, but were elected according to the national gains of their party at the polls. Thus, Mavros' seat was virtually guaranteed before he even entered the race.

97. Papandreou's direct personal involvement in the electoral negotiations with these two personalities is not an exaggeration by any stretch of the imagination. Both Mavros and Glezos by-passed the entire party apparatus of PASOK and negotiated personally with Andreas himself. The personal tone of the public exchange of letters between the three leaders is a case in point. See the Greek dailies 17 September 1981.

98. Glezos, a Lenin award winner and head of the post *metapolitefse* E.D.A., had aquired his reputation as an "honest fighter for democracy" from his extraordinary heroism during War World II in the resistance movement. At this point we should probably note that while Mavros had virtually not lifted a finger to win his seat, Glezos had to run like any other candidate of the Movement to win it. In the end however, he did not have to worry too much, since his prestige and the help of the disillusioned leftists who had decided to vote for PASOK as the lesser of all evils, were enough to put him at the top of the preference list of all Movement's candidates in the tough and prestigious constituency of Athens A'.

99. See Appendix III.

100. The survey was conducted by the Nielson Company and quoted in *Marxistiko Deltio* (Marxist Bulletin). 13-14, 1982.

The Short March to Power Completed

101. *The Greek Economy in Figures 1984*. op. cit.
102. M. Papajannakis. op. cit.

6

The Movement in Power

On 18 October 1981, the Panhellenic Socialist Movement made its entrance to Greek state power with a commanding 48% of the popular vote. Its "short march to power" had come to an end and a new era for both the Movement and the country was starting. PASOK's victory had not only terminated almost a half century of right wing dominance in the country, but the new government was also the first socialist government which the country had ever experienced. For this reason it was greeted with widespread relief and enthusiasm, both in the country and among the left in the West. The Greek Socialists in fact went on to score another victoy in 1985 (with a decline of only 2 percentage points), and these victories emerged in the midst of the right wing euphoria which Reagan and Thatcher had generated. At least momentarily, the Greek Socialists had disproved in a most convincing way, the mythology of a right-wing resurgence.

But enthusiasm is not, as Goethe put it, like a smoked herring which you can preserve for many years to come. As one might have expected, PASOK took with it into government its long standing internal controversies and contradictions and it did not take long for the enthusiasm following the 1981 election to turn first to scepticism and then to disappointment. The new Governmental policies were not only far from fulfilling the inflated expectations of the Greek people, but were not rarely in some contrast to the Movement's previously radical program. Consequently the Movement's leadership had to come to grips not only with the reactions of the population, but also with those of its own party structure, whose enthusiastic activity had in fact brought the Movement to power. However, before we turn to PASOK's organizational reorientation and adjustments to the new circumstances, we will briefly examine the policies of the Greek Socialist's government, which in effect conditioned these latter organizational developments in the Movement's structure.

PASOK in Government

The Economy: From Keynsian Hopes to the Austerity of Despair

When only a few months before the 1981 election PASOK announced its "Contract with the People", there was little disagreement that the document bore only a passing resemblance to the party's 1974 declaration. The earlier declaration's vague but radical strategic goals of "social liberation" as the prerequisite for "national independence", meaningful "democratic procedures" in and outside production and "elimination of the exploitation of man by man" had, as we have seen, been abandoned since 1977. They were now replaced by the goals of "autonomous economic development" and the simultaneous overcoming of "recession and inflation".

For the accomplishment of these goals PASOK put forward a set of neo-Keynesian measures. Production was to be stimulated through raising middle and low incomes, as well as through a "new policy" of incentives for productive investments. Price controls were to be implemented for "basic products and services", in addition to a close scrutiny of public investment and expenditures with a parallel effort to eliminate the widespread tax evasion. "Problematic companies" (companies in debt), were to be rationalized and "structural changes" were to be made in the public sector so that decisive control of the activities of the monopolies would be achieved. For the latter, the "socialization of strategic sectors of the economy" was seen as necessary.[1] Although these "socializations" never went beyond the idea of state control of the pharmaceutical and military sectors, which were in dire straits, PASOK did follow these economic promises quite faithfully. Almost immediately after its victory, Papandreou's government, using the pre-existing corporatist labour practices of the Greek state and a watered down version of its programmatic promise for "Automatic Index Adjustment" (A.T.A.), introduced a wide range of income increases in both the private and public sectors. There was an overall 3% increase in real incomes, which although not sufficient to cover the losses of the previous years (a net real loss of 5.5% in the 1979-1981 period) did amount to real relief, especially for those in the low income brackets.[2]

It was not long before the shortcomings of these "demand-side" economics became apparent. Within a year, Papandreou himself admitted that the Government's policies had not brought the expected results. On the contrary, the economics of the first Greek Socialist Government were creating more problems than they had been intended to solve. The stimulation of demand, which had been undertaken primarily through legislative wage increases, did in fact result in increased demand. This however was not translated into overall increases in domestic production, but rather into increased imports. Domestic producers had tried to offset their higher wage costs by increasing prices. However, due to the already high level of inflation in the country (25-26%) and to the open nature of the Greek economy, this simply tended to encourage people to buy the relatively inexpensive imports. The dependent nature of the Greek economy combined with EEC limits on import controls, served to ensure that the Greek Government could do little to stop this process. In fact the open nature of the Greek economy also tended to retard domestic investment in capital equipment, which might over time have helped to meet the increase in demand without price increases. This is because domestic producers would not have time to recover their capital investments without first being beaten by foreign competitors. Finally, any attempt to meet demand increases by means of increased exploitation was simply not on the agenda. The post-electoral enthusiasm and popular confidence ("PASOK in government, the people in power" – the government's main electoral slogan) hardly constituted the right political atmosphere for wage restraint. In fact, when in some cases, intensification of production was attempted, labour mobilization cut it short very quickly[3] – there were 7.9 million hours "lost" to stikes in 1982, compared to 5.3 in 1981.[4]

Thus the attempt to "stimulate the economy" by manipulating its demand side failed. Investment trends continued to show a negative flow (4.5%).[5] Even the Government's astonishing (given its previous attacks on multinationals) direct political intervention in 1982 to create incentives primarily for foreign capital, did not generate the expected response. Finally, the Government's hopes to exploit the country's (and PASOK's own) traditionally good relations with the Arab world, and in this way to attract investment, proved fruitless. There was a lot of talk and very little action. The international economic situation (recession in the West and a decrease in oil prices) had made not only Arab but also other foreign capital, apprehensive.

After its first year in power, confronted with this situation, the Government began to take a 180° turn – from Keynesian hopes to the austerity of despair. It started to adopt the internationally "fashionable" economic policies associated with monetarism. With these new policies, the much desired recovery was now to be achieved not through the stimulation of demand but through: a) the reorientation of the country's production towards exports and the improvement of the country's competitiveness; b) exchange measures such as the devaluation of the drachma; and c) the reduction of production costs through a wage controls.

The new measures were however more than the mere product of the contradictions of the 1981-1982 policies. They were also a result of the long standing effects which the new international division of labour had had on the dominant commercial faction of Greek capital. Greek shipping capital, which had in the post-war period enjoyed a prominent position in the world's sea transport industry, was reaching its limits both because of emerging protectionism and the world economic recession. Therefore policies which would promote a more competitive, export-oriented resource and manufacturing industry in Greece were in their interests. Such an economic orientation would at least develop a base to compensate for the markets lost abroad, as well as open up new opportunities for their stockpiling of surplus.

However, the continued centrality of the state in economic policy, the country's dependence on foreign loans and EEC subsidies as well as the echo of the Movement's previous (radical) rhetoric meant that this new economic policy would have only moderate results. The output of the primary sector between 1981 and 1985 was on the average a mere 0.9 percent, since savings within that sector financed not rarely consumption and/or the purchase of real estate, or simply were deposited in banks at high interest rates.[6] In the secondary sector, the situation was not any better, as the output of Greek industry, both of consumer and capital goods, remained stagnant between 1981 and 1984. This dismal situation in the "productive" sector of the economy was evident from the employment patterns. A decline of 2.5 per cent per annum from 1982-1985 was witnessed in employment in the manufacturing sector while a simultaneous 3 per cent per annum increase was seen in the already obese public sector.[7]

An even worse indication of the results of PASOK's economic policy can be seen in an examination of the country's fiscal problems. The country's public borrowing increased to reach 17.5 per cent of the GDP in 1985, while the country's foreign debt almost doubled (from $7.9 billion to $14.8 billion in the four year period from 1981 to 1985 to reach an unprecedented $17 billion or 45 per cent of the GDP in 1986.[8] Historically, fiscal deficits were at least partially dealt with by invisible receipts from shipping, immigration remittances and tourism. But the latter two were now in sharp decline having fallen from $13.2 billion in 1980 to $4.6 billion in 1984. At the same time, the interest payments on the foreign debt increased to $1.1 billion in 1984 from $466 million in 1980.[9] Furthermore this situation could not even be much alleviated by the marked increase in exports from 0.3 per cent in 1983 to 5.5 per cent in 1985, since nearly a quarter of the import earnings were used against the foreign debt. And finally, with regard to foreign investment, the Government had only moderate success as it attracted a modest inflow of investments from the EEC countries, and elicited very little response from the much sought after Arab capital. In fact, the Socialists' biggest coup in the field of foreign investment was the economic agreement which they made with the Soviets regarding the exploitation of Greek bauxite reserves, which are the richest in Europe. This was the largest foreign investment ever made in Greece and the terms of the agreement are to an unprecedented degree, beneficial to Greece. As the *Economist* put it, it is a "dream agreement", given the dismal situation in the international aluminum market.

The rather blue state of affairs in the economy was in fact to strengthen the technocratic voices within the party and the Government. Thus after a short interlude of high spending just before the 2 June election, which along with the "anti-Karamanlis syndrome"[10] proved enough to ensure PASOK's re-election, the

The Movement in Power

Movement completed its 180° turn away from its original Keynesian orientation. The cynical rationality of its technocrats who started to become increasingly vocal,[11] pushed aside the post-electoral weakness of the *palaiokommatikoi* and the timid, abstract intentions of the almost marginalized left-wingers and became Government policy. In October 1985, Papandreou's Government announced an austerity package for the "stabilization" of the economy. This included a 15 per cent drachma devaluation (the second since 1981), an open door to foreign investment, and generally the promotion of incentives for private capital with some deregulation in the legal and tax framework in the art of profit making, cutbacks in social spending (Greece in fact has the lowest expenditure on health and education as a percentage of GNP of all the EEC countries), further relaxation of price controls, greater flexibility for employers not only with regard to hiring and firing, but also with regard to the use of labour (e.g. working hours), and last but not least, a reverse indexation of wages. The latter in fact appears as the corner stone of this austerity programme. The previously legislated "Automatic Index Adjustment" now meant that wage increases had to correspond to the targeted imported rate of inflation and not to the actual one. It is obvious that the back bone of the "stabilization program" is the lowering of labour costs, which will assist the promotion of low and medium technological range exports and tourism. These austerity measures, which have been accompanied by anti-strike actions, and an ever stronger emphasis an "social peace", can easily be said to have brought about the desired results since in 1986 alone, (and always according to official estimates), wage earners suffered an 8.6 per cent loss in real wages. However, this "attack" on labour has not been a constant in the Socialists' labour policies.

Controversial Labour Policies

The Socialists inherited a markedly corporatist labour relations system in which compulsory government mediation, and other direct involvement in trade union procedures were the rule, and free collective bargaining and open/democratic trade union structures the exception. PASOK did not appear to be disturbed by this heritage and carried on the customary un-democratic practices as it proceeded with the arbitrary (using some legalistic excuse) replacement of the leadership of the National Federation of Labour with party faithfuls. Papandreou's government hoped that it would not only be able to extend its influence among the working class, but also to mobilize support for its labour policies.

During the first year in Government, PASOK did not in fact need to exert much extra effort to legitimize its labour policies. In addition to the wage and pension increase which came as part and parcel of its initial Keynesian policies, the Government put forward a number of laws which, at the time, were justifiably considered pro-labour. There is no doubt that some of these measures, despite the patronizing rhetoric which came with them, were unprecedented. Lockouts became illegal, the possibility of the "legal" prohibition of strikes was eliminated, labour's right to organize was solidly protected, and finally a number of scarcely pre-existing labour benefits (e.g. vacation, pensions, maternity leave, unemployment insurance, etc.) became legally guaranteed. The legitimacy lent to the Government in the eyes of the unions was bolstered even further by the effective support which both Communist Parties provided to the new Government.

By mid-1983 however, the Government's move towards austerity, and the crescendo to austerity since October 1985 came to be accompanied by diametrically different labour policies. The Government decided to abandon its policies of inclusion and concensus and pursue policies of exclusion and coercion. Rather cynically pointing to the real problem of low productivity in the public sector, the Socialists introduced a law which in effect banned the right to strike in the public sector. The infamous "Article 4" of this act, which ironically was called "on socializations", placed a number of obstacles in the path of the militant initiatives of labour. For example, in order to become "legal", a strike vote had to be 50 plus one per cent of the registered membership and even if this highly unlikely number was achieved, the decision had to be approved by the tripartite administration (i.e. the state, the board of directors, union representatives) of the perspective organization or public enterprise. This law was of the kind, as many liberal observers in Europe noted, which not even had Mrs. Thatcher dared introduce in her first term.

In spite of appearances, this dimension of the Socialists'

labour policies was neither "contradictory" nor "confusing," nor a "flip-flop" between pro and anti labour initiatives, as the opposition maintained. On the contrary the new labour strategy in the public sector was both consistent with the Government's overall economic policies, and complementary to the previous introduction of liberal legislation for the private sector labour relations. A close look at the basic traits of the Greek state sector is enough to prove not only the inadequacy of the opposition's criticisms but also to help us understand the controversy of the Socialists' labour relations strategy.

Two important characteristics of the Greek public sector will serve to underline the significance of this move. First of all, as we have seen in the first chapter, the public sector in Greece extends well beyond the usual activities of western capitalism. It encompasses almost 90% of the financial sector, all land and air transport, all communications and even some segments of heavy industry. Second but no less important is the fact that the public sector represents the most radical expression of trade unionism in the country. Consequently, qualititavely different labour relations in this sector could in effect set the limits and the pace of labour demands throughout the entire country. Furthermore, since the sector is very much tied to the activities of private capital, such new "labour relations" would improve the "investment climate". Keeping this in mind it is rather difficult to sustain the argument of confused and contradictory labour relations. The desire dictated by "supply-side" economics for severe reduction of labour costs had to be achieved, at least until the 1985 election, with the least possible political cost. In fact this is exactly what the Government was trying to do in the most skillful way possible.

Thus in the private sector, wage concessions were to be left to the market mechanism, which, given the growing threat of unemployment, worked effectively as a disciplinary device. In fact, the Government (very much contrary to other western governments), not only does not miss any opportunity to advertise the rate of unemployment but it seems that the various departments and the prime minister himself "compete" as to which can quote a higher rate.[12] This tactic has been rather successful, if we can judge from the reduction of labour unrest (working hours lost to strikes were down to 2.9 million in 1983 and 2.7 million in 1984 from 7.9 in 1982[13]), as well as from the open wage concessions actually secured in the private sector (particularly in businesses which were under the special "rescue plan" of the Government).

In the public sector, however, where the unions are strong and tenure is granted to employees as a long standing practice, the Government chose a head-on collision with labour. Moreover, using the otherwise reasonable argument of striving for "a more efficient public sector", the government's efforts to legitimize its direct disciplinary actions against public sector workers had devastatingly divisive effects on the labour movement. Those frustrated private sector workers who had been laid off and those young university graduates who have never been employed have not only come to believe in the Government's initiatives but have also gradually started to see "Privileged" civil servants as the enemy. It is not difficult to imagine what paralyzing effects such conditions have had on labour movement militancy, especially since this objective bitter reality is combined with the incapacity of the left opposition to put the Government's labour policies in perspective and oppose them effectively.

Thus PASOK's labour policies are of a piece with the Government's overall economic policies. Using the well known tactic of the carrot and the stick for private and public sector employees respectively, Papandreou's Government was trying to discipline labour and create an "environment of social peace" and a "moderate political climate".[14] However, since 1985, the Government no longer leaves the project of lowering labour costs to unemployment (which has jumped from 8 per cent in 1984 to 10 per cent in 1987, and out of which, 60 per cent of the unemployed are under the age of 24) and to the divisiveness in the labour movement. It has taken more aggressive stands such as back to work legislation, the use of the army to break strikes and constant threats of legislation which will stop the "abuse of labour privileges" have become common themes on the every day political agenda.

This attack on labour has not passed without comment or political cost for the government. The trade union movement (one third of the labour force) has organized a number of rallies and strikes, in protest against this government policy since 1985. These reactions did not leave a number of militant trade unionists within PASOK unaffected. In fact, after failing to change the Government's policy, a number of them withdrew from PASOK and PASKE, and under the banner of the newly established trade union organization the Socialist Workers Movement (SSEK), they organized against the government within the labour movement. However, with the right in tacit support of the austerity programme, and the left incapable not only of opposing the Government, but also of coming up with concrete positive proposals for a way out of the crisis, overall the Socialists' labour and economic policies have not been under any political threat. Ironically, the only critical questioning of these policies came from within PASOK, namely from the *palaiokommatikoi*. Knowing very well that in the next election it will be difficult to rally support for these unpopular economic measures, the *palaiokommatikoi* have been questioning the technocrats insistence on the "stabilization programme". It is therefore to be expected that as we approach the next election (officially the summer of 1989) that the voices and pressure of the *palaiokommatikoi* will become more effective and convincing.

A Unique Foreign Policy

We have seen that PASOK, while in opposition, as a result of its understanding of the country's dependent political economy, had adopted a certain rhetoric regarding the country's foreign policy. More consistently than in any other aspect of its politics, PASOK had insisted that the US, NATO, and the "West" were the causes of the "Greek tragedy" and had put the struggle against these forces at the top of its political priorities. Although in its 1981 governmental program, the high-pitched tone of its anti-imperialist policies (i.e. oust the American bases, withdraw the country from NATO, cancel its membership in the EEC) had been considerably softened, there was still a strong sense that many of these promises would be implemented.

But again it did not take long before the Socialist foreign policy disappointed these expectations as well. Although maintaining part of his radical rhetoric, Papandreou has managed to undertake policies diametrically opposed to those contained in his party's original plans. In the summer of 1983, the Socialists reached an agreement with the Americans which in spite of the government's claims, extends and guarantees the presence of

US military bases until the end of 1988.[15] Meanwhile, all observers agree that the renegotiations which started in September 1987 for the extension of the agreement will conclude a new agreement on a more permanent basis. That is likely to be so even if Papandreou has to ratify the agreement through a referendum, as he mentioned. In fact, in this case, given that the result will be positive, Greek-US relations will gain the much needed popular legitimacy, and the objections of the left will be completely neutralized. Furthermore, in exchange for the maintenance of the 7/10 ratio of US military assistance to Greece and Turkey, PASOK's government will continue to actively respect the Rogers Agreement which fully reintegrates the country into NATO's strategic plans in south-east Europe. The highlight of this turnaround in PASOK's foreign policy is the decision to purchase an enormous amount of military equipment (100 aircraft – of which 70 will come from the US and the rest from France – tanks and sophisticated surveillance devices). This sale (which the Greek press hailed as "the purchase of the century") not only perpetuates the country's military dependence upon the US and NATO, but, due to its enormous cost (Greece already spends over 7 per cent of its GNP on military expenditures), also opens the gate for very unpleasant financial consequences for the country.

There is no doubt that Papandreou's "anti-American", "anti-NATO" pronouncements generated a hostile attitude from the Reagan administration and it is clear that great pressure has been applied to the Greek Socialists. But the Government has too often gone overboard to accommodate US interests – for example the voluntary inclusion of the demilitarized parts of the country into NATO defence plans, the acceptance of US operated AWACS, the expansion of NATO installations in Macedonia and Corfu, or even the provocative re-erection of Truman's statue in down-town Athens on the anniversary of the bombing of Hiroshima (1987) in total contempt of popular feeling. Since 1985, the new element in the Socialist's relation to the US and NATO besides the regular visits of American officials to Athens and the presence of the US fleet in Pireas (something which had ended along with the colonels' regime) is the fact that Papandreou links relations to NATO and the US with the country's uneasy relations with Turkey. NATO and the US are increasingly seen as guarantors of peace in the Agean. The latter is in fact the basis for the expansion of popular consent for the

Government's initiatives *vis-a-vis* the US and NATO.

Given all of the above, it is only fair to claim that the Socialists' foreign policy is not much different from that of its right-wing predecessors. In addition to defence policy, it did not take PASOK long to realize the oversimplifications contained in its previous analysis of the EEC. Thus, not only has the country's loyal membership in the Community continued, but it has also taken full advantage of its grants, particularly those given to the primary sector as well as the recently introduced Mediterranean Integrated Programmes (MIP).

At the same time it is also important to note that to its credit, the PASOK government has not fallen into the fashionable cold war rhetoric of the other socialist parties in Europe. For example, Papandreou's anti-nuclear rhetoric and his active participation in the "peace initiative of the six" – though from the outset condemned to failure – has undoubtedly contributed to the dilution of the cold war atmosphere on the continent. PASOK has also been a supporter – although a suspiciously uncritical one – of the P.L.O. and the Arab countries, particularly those with so-called militant governments (i.e. Libya, Syria, Iraq). Finally, much to the dismay of the short-sighted Reagan administration, the Socialists have maintained good relations with the Soviets and the other Eastern European countries. Papandreou refused to sign the EEC's cold war document which condemned the Soviet Union following the Korean Airline incident; it mediated in the negotiations for transforming one of the private shipyards into a servicing station for the Soviet fleet in the Mediterranean; it has been provocatively silent on the Afghanistan issue; and finally it has even gone so far as to be a very strong supporter of the Jaruselski regime, calling it "the best thing that ever happened to Poland".[16]

However, if we are to understand the material base and the limitations of such controversial policies, some light must be thrown on the motivating forces behind the Government's initiatives and silences in the formation of the country's foreign policy. Papandreou's good relations with, and open support for, the countries of the Soviet block must be attributed to: a) Greece's traditionally good (though somewhat on the decline) trade relations with the Comecon countries (almost 10% of Greece's exports), which even under the Junta remained strong; b) the growing interest of Greek shipping capital during the last twenty years in expanding its activities in these countries;[17] c) the effort both to attract Soviet investment (as the case of the Greek bauxite described above) as well as to accommodate the ventures of other Greek capital, such as the recent undertaking of huge projects in the Soviet Union by Greek construction companies. Finally, and probably most importantly, these policies must be seen as an effort to neutralize any effective opposition from the C.P. which dominates the political space to the left of PASOK. So far, the practice has in fact been successful, if we can judge from the C.P.'s habit of taming its response to strike activities whenever they coincide with positive Governmental gestures to Soviet policies. Still, these reservations about PASOK's foreign policy should not lead to a dismissal of its positive aspects. Greek socialists are undoubtedly an oasis in the European desert of cold war, and this should not be forgotten.

The Socialist Legislative Record on Other Fronts

There is no doubt that the Greek Socialists have introduced a number of positive reforms and progressive social programs in addition to those cited above with regard to labour. As examples, we can cite the liberal reforms in family law and in the legal framework regarding the overall position of women, which although watered down in their final version due to church pressures, are definitely a step forward, so far as the position of women in Greek society is concerned. The introduction of civil marriage, the relaxation of divorce laws, the abolition of the archaic laws on dowry and marital contracts and the legal recognition of the right of women to enter into a contract without the consent of their husbands are some of the most significant reforms in this area. In addition, the Government's references to equal pay for work of equal value, although far from being translated into reality, are promising steps in the right direction.[18]

In addition, the Papandreou Government confronted the opposition of the academic establishment and introduced extensive educational reforms (especially in post-secondary institutions) which, both in form and in content, opened the doors for greater participation on the part of both students and junior staff. Similarly, brushing aside the threats of the medical establishment, the Socialists are actively trying to fight their way through the countless private medical plans and are working on the introduction of universal medical and pension coverage. At the level of administrative reforms, the Government came forward with the implementation of its longstanding promise of political decentralization and extended the juristiction of local and municipal authorities, thus creating a favourable legal framework for mass participation. (It is of course not surprising that due to the country's undemocratic political tradition, which was anything but conducive to mass participation, such participation is far from evident up to this point.)

Finally, among the Socialists' positive measures has been the recognition of the War World II Resistance Movement, as well the compensation of the veterans involved. This latter may not appear so important to outsiders but there is no doubt that this amounted to a strategic slap in the face to the considerable anti-communist forces in the country. In fact in light of the strength of the anti-communist discourse in Greece from the time of the civil war, PASOK, must be credited with having re-established the legitimacy of socialist language in the country's political discourse. However, as we have seen above, this is not entirely an unmixed blessing.

Strategy?

Given the mixed and controversial nature of the Government's policies, it may appear that PASOK is engaging in nothing more than a strategy of hit and miss pragmatism. This is however, not the case, since as we have seen, the Movement is hardly a homogeneous political organization. Of course, after 1981 and its controversial organizational developments, PASOK was less diverse than ever before. However, its initial tri-polar political base made up of the technocrats, the *palaiokommatikoi*, and the left (though now small and bureaucratized) did not disappear. On the contrary, as one might have expected, the tendencies of the organization moved into the governmental apparatus and began translating their differences into Government policies.

In fact, the distribution of the cabinet portfolios by Papandreou, who of course remained the connecting power pole for all these currents, revealed not only the existence of these currents, but also gave away the actual balance of power among them. The majority of these portfolios were distributed among the technocrats and the *palaiokommatikoi*. The technocrats were consistently given control of all the departments having to do with the economy from National Economy to Energy and Research and Technology, while it appears that the *palaiokommatikoi* felt more comfortable being placed in command of the state administration, as they were given control of portfolios such as Internal Affairs and Security. And finally, the weak remnants of the radical tendency were given third class portfolios, although admittedly with symbolic value, including such portfolios as the Ministry of Youth.

Thus objectively the Government has become the arena of the Movement's internal politics. This is the key to understanding the changes in the Socialists' policies and in detecting the strategy behind them. A grasp of the political orientation and the social base of these tendencies should assist in the comprehension of the diversity and the confusion of the Government's policies.

There is no doubt that each of the tendencies has a more or less clear political strategy. For the technocrats, who have been numerically powerful in the Movement since 1977, and politically so since PASOK gained power, the Greek problem was quite clear. Greece had to overcome its "parasitic" connection with the international economic realities and "make up the time lost due to right-wing mismanagement."[19] The country would be able to survive and expand only if it was able to adapt to the new international economic circumstances – it would be necessary for it to abandon its old place as intermediary tradesman and become involved in more productive activities.[20] This need for rising productivity and restructuring of the economy along industrial lines was an explicit position stated by various proponents of the technocratic tendency particularly after PASOK achieved power. As G. Arsenes, former director of the Bank of Greece and later Minister of National Economy who in 1985 quit the party over the intensity of austerity measures put it, after having indicated the decline in capital accumulation (from 23% in 1971 to 17% in 1980), *"allage* means accumulation".[21] The same type of message can be seen in a number of official Government statements, especially in contacts with the business world. In fact, Papandreou himself, who favours the technocratic strategy, of course within the limits of his position as leader of the party and the Government, spoke about this a few months after forming the Government. He told the Congress of Greek Industrialists that "in Greece, the dominant class is not the industrialist but the middle-man,"[22] which was widely interpreted to mean that it was the latter of these two categories which was now the problem.

These strategic goals of the technocrats were in fact crystallized in some particular Government policies. These included unprecedented incentives to the private sector, planning the concentration of production in the primary sector (with the assistance of the EEC), the restructuring of public enterprises and the strengthening and modernizing of the banking system so that they could assist the "expansion of domestic companies and the attraction of joint venture partners...",[23] the attempts to control the up to that point unscrutinized activities of the middlemen through taxation, and the attempts to control the traditionally high activity in construction through the introduction of property tax. It was statements and economic policies such as these that have contributed to the eventual development of smooth relations between Greek industrial capital and the Socialist government.

The *palaiokommatikoi* on the other hand, apparently had no comprehensive strategy nor even a vision as to where the government should move. However, being politically the most experienced within PASOK's organization and having strong support within the countryside among the farmers and in the cities among the petty-bourgeoisie, the *palaiokommatikoi* were far from ineffectual in the policy formation of the Government. They had basically left the policy initiative to the experts in the Government (i.e. the technocrats) and their primary concern was the delivery of the goods to their constituents so as to ensure their re-election. This meant the development of an efficient state apparatus capable of undertaking the much needed social services (something which was to unite them at least in part with the technocrats) and the protection of the necessary conditions for the reproduction of the social strata which depended on small ownership. Thus it seems likely that it was pressure from the *paliokommatikoi* in cabinet which resulted in the first Socialist policies being of Keynesian and not monetarist form. Social benefits to farmers, tax breaks for low income earners, and increases in pensions were prompted by the *palaikom-*

matikoi. In addition, it was the *palaiokommatikoi*, with the interests of their small propertied constituents at heart, who very actively opposed the Department of Finance decision to introduce a property tax. They applied considerable pressure to Papandreou who became so nervous about this internal and the external reaction that he withdrew the bill on national radio without so much as consulting the minister responsible.[24]

However, ultimately, it was the inescapable forces of the market which seemed to resolve the tensions between the technocrats and the *palaiokommatikoi*. As we saw, the first year plan of the Keyensian road to economic management and development failed, and the more crude approach of monetarism was soon to become more and more the basis of the Government's policies. The poor showing of the economy during the first year of Socialist rule meant that the *palaiokommatikoi* concerns had to be put aside if the "national mission" of restructuring the economy was to be successful. The goal and the intentions are clear: "*allage* now indeed means accumulation" and higher productivity, which are to be achieved by any means. In the fall of 1985, Socialist economic measures which introduced drastic wage restraints and the cancellation of even the existing free collective bargaining procedures indicate that the technocrats have, thanks to overall economic conditions, finally won the upper hand among the Movement's tendencies. However, again, the unpopularity of the "stabilization programme" has strengthened the voices of the *palaiokommatikoi* within the government since the summer of 1987. These voices are in fact likely to become stronger and more effective in changing or modifying the technocratic project as the election approaches. Thus the predominance of the technocrats should not be taken to imply the terminal marginalization of the *palaiokommatikoi* within the Government. On the contrary, the *palaiokommatikoi*, thanks to their political skills are useful in order to win elections, and of course between elections, when they can skillfully battle the opposition and make the Government's plans a saleable commodity. These are skills which the technocrats themselves not only are lacking but also, given their limited ties with the electorate, are having a difficult time dealing with. It is obvious from this configuration, that although for the moment the technocrats are clearly in command of the Government, the balance of power between the two tendencies is delicate. The continuation of this equilibrium will depend on the state of the

The Movement in Power

economy and even more importantly on how much, and for how long, the working class and the social strata based on small ownership economic relations, are able to to put up with wage restraints and with restraints on their reproduction.

Until 1981, the left tendency continued to occupy certain locations within the organization's structure, holding positions of importance among the Movement's youth and at the local level. Of course, after 1977, it lost much of its dynamism and its closeness to PASOK's rank and file, and over time, the leadership of the radicals was in fact fairly well incorporated into the party's establishment. Within an organization incapable of articulating political debates or the exchange of ideas, they became bureaucratized and lost much of the power they used to have. The presence of the "left" in the Movement and eventually in the Government was actually rather useful as it was often used as an alibi for questionable Government initiatives. This is in fact where the power of the radical tendency, (*vis-a-vis* the other tendencies) lay. For this reason, the left was given special ministerial portfolios through which its presence in the cabinet could be advertised, yet through which, it could exercise very little power, since its jurisdiction had been so carefully ghettoized. Under these conditions, the contribution of the left to Government policies could only be limited and symbolic, and the tendency became incapable of modifying in any substantial way, the overall direction of the Government. PASOK's left tendency can however be seen as responsible for that part of the Government's record which was essentially symbolic in nature and which carried no heavy cost burden. This includes the recognition of the resistance and the reforms in family legislation. The left's inability however, to even modify the austerity measures after the autumn of 1985 seems to have broken this current's limits of political toleration. Under the pressure from their constituencies, especially in the trade union movement, they withdrew from the Government and the party, and undermined in this way one of the three pillars of PASOK's rise to and maintenance of power. Thus the absence of the left (as a alibi of course) within the Movement is likely to have devastating effects on the party's and the Government's future.

To sum, a careful consideration of the main thrust of the Greek socialists' policies, in light of the power distribution among the three tendencies, can hardly lead to the conclusion that these policies are incoherent, or that they lack strategic goals. It can also not be argued that the policies of the Government are the mere outcome of a full-blown struggle within the state apparatus between these three tendencies. For the time being, it seems that the dictates of the omnipotent market have solved the problem by awarding the upper hand to the technocrats. The often contradictory appearances of the Government's policies are not the product of administrative or other confusion, but are rather the result of the influence of the *palaiokommatikoi* and to some degree (until 1986), the impact of the marginalized radical tendency. PASOK's policies represent the attempt of Greek capital to overcome its parasitic position, to restructure, to become productive and as such to jump on the wagon of the new international division of labour. In this sense, Papandreou's government has become the modernizing Jacobin of the Greek social formation.

The Socialist State: Harvesting the Results of PASOK's Party Structure

If there is one area in which the PASOK Government has surpassed the fears of even its most bitter critics, it is in the pattern of state administration. PASOK's presence in the edifice of Greek state power has promoted a general tendency toward authoritarianism in the central state apparatus – very similar to the one described by Poulantzas in his last book *State Power and Socialism*. Political parties, as we have said, are portraits in miniature of the political project which they are to carry out once in power, and PASOK is no exception. After the above extensive examination of PASOK's development, however, the Socialists' record on this front should not come as a surprise.

In spite of the gestures and rhetoric indicating otherwise, since 1981, we have witnessed an unprecedented centralization of state power. The new decentralization programme which was to give more power to the local governments remains an empty letter of the law. Effective power still lies in the top central state apparatus which has become, if anything, even more distant from popular control. Parliament has faded constantly into the background of the configuration of the Greek state, and immense power is being concentrated in the Prime Minister's hands, a situation which was exacerbated by the 1985 constitiutional reforms. PASOK's backbench M.P.s have actually been prohibited from asking questions in the House and Papandreou has made a habit of making himself scarce in parliament. Appointed in effect by Papandreou himself, members of PASOK's Executive Bureau decide on major Governmental policies, undermining in this way the traditional functions of both the party's parliamentary caucus and the cabinet.[25] In this way, the Socialists' configuration of state power tends to be concentrated more and more in the hands of one man – A. Papandreou.[26] Far more ominous however has been the Socialists' continued exclusion of certain sectors of the Governmental apparatus from democratic control. In the name of "efficiency" and "neutrality", PASOK has systematically exempted the military, the diplomatic service and the justice department from parliamentary or public scrutiny. This is especially troubling since Papandreou's Government has surpassed its immediate predecessor on the scale of state coercion practiced in certain areas, as the repressive state apparatus has been considerably strengthened. The heavy-handed policing of strike sites, the increased surveillance of youth hang-outs, the crackdown on a small number of punk youth and the constant harassment of gays, are some striking examples of this depressing picture.

Furthermore, the Socialists have added a new dimension to the Greek political culture – the technocratic definition of politics. Traditionally orthodox governmental efforts to fill bureaucratic positions with the party faithful have now been accompanied by a rationale for patronage appointments which stresses the unique "expertise and competency" of the appointees who can by definition deal effectively with the country's problems. Party loyalty, while important but secondary in the high positions, is a must for the middle and low range bureaucrats. The phenomenon of "green guards" – as the inventive sarcasm of Greeks named it – has taken immense dimensions. These patronage appointments (based on party membership) may very well be understood as a continuation of the long tradition of clientelism. Now however, this pattern has been articulated in the instrumentalist arguments of PASOK's political discourse, as the occupation of civil service positions by party members became by definition, the guarantee of the success and the proof of the correctness of the Government's policies.

Moreover, during the first few years of the Socialist administration we have witnessed a further submission of the institutions of civil society to the policies and the logic of state activity. The systematization of corporatist practices in the trade union, cooperative and students' movements are living examples of this tendency. Every expression of civil society (political, cultural or otherwise) in the last few years has been pushed towards expression only within the strict framework of the party system's logic and the centralized practices recognized by it. This is the case not only when independent political or cultural initiatives want to gain legitimacy but also if they do not want to find themselves excluded by coercion from the terrain of their activity.[27] Of course this is not surprising, since the Greek state has a long history of non liberal practices. In addition, as we have seen, the state in the Greek social formation carries a particular weight in the articulation of social relations. It is the state and not civil society which has been proven the arena *par excellence* in which the social classes articulate their presence and secure their reproduction. For this reason, the main thrust of the practices described above is neither unusual nor totally new. What is new however, is the political discourse which accompanies the state's actions under the new Socialist rule.

To begin with the intense anti-imperialist/anti-monopoly rhetoric often takes simplistic and anthropomorphic forms as it is expressed in Manachean and chauvinist terms. It never reaches the point of becoming anti-capitalist, especially when confronted with the country's socio-economic problems. This can be seen in the delicate "eclecticism" of the government which allows extensive reporting on the British miners' strike or even mobilizes solidarity for Nicaragua but can (and does) without any political risk, make a habit of hushing up any local militant activity. So far as the practical applications of this part of governmental discourse are concerned, they are at best exhausted with propositions for a redistribution of incomes and rationalizations characteristic of a "truly" welfare state.

Second, there is a constant and emotional attack on the pre-existing corrupt "establishment", which it is said, can be changed by the systematic promotion of the ideology of "participation". In the reality of power however, PASOK has translated its initially loose definition of participation to mean the participation of experts in the state apparatus. It is this technocratic definition of participation which not only allows the agents of the new middle class (i.e. highly educated and specialized professionals) to take a step into the chambers of power, but also contributes to the decomposition of the social strata which were traditionally nourished by right wing clientelism on the fringes of the state structure. But even when the materialization of this "ideology of participation" is attempted in terms of actual mass involvement in state policies, as in the 1983 campaign entitled "PASOK Talks with You – Dialogue With the People", the whole effort was proven to be a mere facade and a pompous public relations enterprise for the Government's policies.[28]

Finally, the Socialist policies are accompanied by a rhetoric which promises a "rational solution to problems", which in turn is to be achieved through "meritocracy" and the promotion of "fair chances in all spheres of public life". Thus, all the complaints about "inequalities" cannot develop fully into a protest which will touch the structural causes of the problems. Every evil is now understood as a product of mismanagement or incompetence, and all sicknesses are magically transformed into symptoms. The Socialists in power have adopted an a-historical fetishistic attitude *vis-a-vis* state power.

The Organization: Dealing with the Fruits of Hyperactivity

In October 1981, the Movement's organization had just carried out a successful electoral campaign. Its strengthened membership, after long and painful preparation had just finished their short march to power, inaugurating not only a new era for the country's politics but for their own organization as well. After being in opposition for more than eight years, the organization had to learn to adjust to the intricacies of a government party. Thus, it is no surprise that in the period under examination the Movement's organization was preoccupied with one project: that of adjusting to the new conditions.

PASOK's victory and the termination of the over forty year old rule of the right, meant in addition to anything else, an urgent need for new recruits to populate the administrative apparatus. Naturally therefore the new government had to turn to its own organization in order to fill this need. Three to four thousand activists of the Movement's organization, according to an unverifiable source,[29] were recruited to fill the various Governmental and administrative positions. The extent of this recruitment was heightened by the fact that all members of the Executive Bureau entered cabinet holding major portfolios. This, in combination with the electoral orientation of the organization, and its undemocratic and centralized structure (which was anything but conducive to the training and production of new activists capable of carrying on the organizational tasks of the Movement), had devastating effects on the organization. The result was that the highly energetic organization fell into deep hibernation. That condition was more evident among the local clubs especially in the countryside and to a lesser degree among the professional ones, since it was the former which, more than any other part of the organization, had an electoral orientation.

The dynamic enthusiasm of the organizaton, which had been so effective electorally, obviously could not disappear overnight. In addition, the overall radical tone of the Movement's policies, the promise for mass democratic participation, along with the instrumentalist premise in its policies, created a legitimate framework for some of the most lively parts of the organization to attempt to intervene in the policy formation of the new government. These sections of the organization were none other than the professional clubs or the *kladikes* which soon became notorious in the country. This part of the Movement's organization was not only by definition less oriented towards electoral activity but was also composed of some of the most politically astute members of the Movement. In addition, given the technocratic orientation of the government, they became the platform *par excellence* for the Movement's input into governmental policies. Moreover, the fact that most of the Movement's recruits into the government came from the *kladikes* gave to these energetic members the needed legitimacy to intervene in the government's policy formation and police the observance of "the contract with the people". But, given that the Movement had no comprehensive and concrete plan for the implementation of its programme, nor had it made explicit its actual intentions on the question of party-government relations, nor had it established a mechanism of dialogue and exchange for developing answers to these questions, the result of interference of the *kladikes* could be nothing other than anarchy. The *kladikes'* activists, soon to be named "green guards," flooded the governmental apparatus at all levels, as clerks and political advisors of various departments, and as directors and researchers of state enterprises. They became the policing voice of the Movement in the government. The heterogeneous social and political origin of these activists, which had all along led them to read their own political definitions of the organization's socio-political project into the Movement's vague programme and promises, made their interference particularly disfunctional to the government.

Even more, the "green guard" activity threatened the overall unity of the government, which was composed of and meticulously balanced between the two major remaining currents of the Movement: the *palaiokommatikoi* and the technocrats. On the one hand the *palaiokommatikoi*, having a natural distrust for the Movement's extra-parliamentary organization saw their interference as an attempt gain submission and control through a non-elective mechanism and thus violate their independence as elected representatives of the nation. In fact this was the main reason that led some of them to react rather drastically.[30] On the other hand, the a-political technocrats could not function with any number of overly politized members breathing down their necks. Thus, the image of the Greek socialists in their first couple of years in power was essentially one of chaos. But this was hardly surprising since PASOK's entrance into power was bound to be accompanied by the transfer of its internal antitheses into their exposition within the echelons of power.

Confronted with this situation, the Movement's leadership, which in any case – as the short history of the party vividly displayed – was not used to having any restrictions placed on its political flexibility, and understanding the political consequences, decided to stop it. The flying enthusiasm of this part of the membership, and its radical implications had to be brought down and under control, even if the landing was going to be a rough one. The first response to the increasingly uneasy relationship between the party and the organization came from Papandreou himself as early as February 1982, when he identified "the phenomena of disfunctional and unfavourable (public) impressions which the right has tried to exploit accordingly". The organization, Papandreou continued, is "the consciousness of the government" and should not act "...as policeman to the Government"[31]. However, the advice and the general guidelines of the leadership to the membership were not sufficient to control the Movement's post-electoral enthusiasm.

In fact the presence of, and the pressure from, the most radical elements of the Movement's recruits in the governmental apparatus grew disfunctional, as the Government started to move to more open compromises with the "establishment". These compromises ranged from unprecedented incentives to the private sector (Act 1262/82), to the recruitment into key positions – especially in state corporations – of technocrats with ambivalent political backgrounds. This was not because the expressed criticisms towards the government had taken any polemic or destructive form. In fact quite the opposite was true, but the problem arose because the Movement, even in opposition, had not developed an internal mechanism of critical dialogue, and because the leadership had managed to establish organizational patterns which were anything but tolerant to such rank and file criticisms. Any further intensification of the rank and file's criticisms and direct pressures would have resulted in a severe restriction of the Government's flexibility which was now more necessary than ever before, as the Government was confronted with the hard reality of power. Thus, to the leadership, this danger had to be stopped in an immediate and effective way. Since the internal memos and seminars[32] did not work, the organization's well tested methods of control had to be put in action.

Consequently, in the first year-and-a-half of the Socialist administration we once again witnessed the familiar phenomena of coercive action on the part of the leadership towards sections of the organization which insisted on acting in contrast to the advised guidelines from the top leadership. In the cases in which the criticism of and the intervention into governmental affairs by the "laymen" of the organization got out of control, the organization reacted violently.[33] Thus, the familiar phenomena of dismissals of elected bodies of the organization, the administrative interference of the leadership in the life of the "undisciplined" local and especially the professional clubs, and even the actual abolition of some of them, were not uncommon.

Coercive action was not however the only means used by Movement's leadership to tame the membership. Coercion without concensus can be dangerous, and this was something that the leadership could not afford. To gain this consensus, the leadership knew from experience that it could not simply rely on vague organizational guidelines and abstract theoretical statements for the unsophisticated and untrained membership to follow but rather that it needed some more material initiatives. These were none other than the co-optation of the Movement's organization through massive recruitment into the state apparatus. The essentially instrumentalist conception of the state and government became objectively the platform *par excellence* for the leadership's response. The entrance of PASOK's membership was to be the living proof of the actual input of the Movement's organization into the Government's policies. In this way the membership could be fetishized and remain silent, as in a "real" and concrete way it was part of the grand administrative apparatus.

It was consequently during this period, that the Greek state experienced possibly the most extensive patronage recruitment in its history. An army of members, especially from the Movement's *kladikes* (professsional organizations), entered the governmental machine at all levels. Such was the eagerness of the leadership to incorporate, and subsequently to neutralize its own organization, that membership in one of the Movement's *kladikes* became the guarantee for a job offer and/or of job security. This explains the fact that in spite of the overall post-electoral inactivity and disorientation of PASOK's organization, its membership, especially among the *kladikes* expanded geometrically, from 110,000 to 200,000.[34] As the unexportable unemployment of the country was reaching unprecedented levels, these patronage appointments soon took extreme dimensions as the huge public sector was recruiting exclusively from within the Movement's ranks. According to at least one source, the Movement's membership was close to 250,000[35] by 1984, out of which 70 per cent were post-1981 recruits. Furthermore, 89 per cent of this new membership had some kind of professional or occupational link with the state apparatus.[36] In fact, despite PASOK's public denials on the extent of the patronage appointments, the Movement had to systematize the pattern by transforming its "Solidarity Bureau" into a man-power office.[37]

There is another very important dimension to the PASOK leadership attempt to control the organization. As the Government was composed primarily of representatives of the Movement's *palaiokommatikoi* and technocratic tendencies, the need to maintain the delicate balance between them now became even more critical than ever before. The Government cannot afford the political cost of further disruptions to its unity, which, as the first year's resignations from the government and the parliamentary caucus indicated, could take uncontrollable dimensions. Thus, patronage recruitments can in a peculiar fashion, function on the one hand as a balancing factor between these two currents, and on the other as a bridge for the traditionally uneasy relationship between them and the party's structure. The *palaiokommatikoi* of course, had a "natural" attraction towards clientelism and now they loved it even more as the Movement's systematic approach depersonalized it and made it less unpopular. For the technocrats, this pattern of recruitment was at least tolerable since in spite of its partisan base, it was claimed to be an apolitical way of conducting politics and even further under its current expression, it had technocratic overtones. After all, it was argued that the recruitment was taking place on the basis of merit, since it was done almost exclusively through the specialized professional clubs. Thus the new version of the old clientelistic practices served as a safe and politically legitimate alibi for the technocrats' a-politicism. Finally, the left activists, whose major concerns, at least until 1986 when the majority left the party, were policing the *allage* and patrolling the "third road to socialism, perceived the patronage appointments as a means of influencing Government policy. In this way ironically, these clientelistic practices of the Government became in effect a means of coopting the most radical part of the party.

The Organization as an Appendage to the Government

The response of PASOK's leadership to the organization's pressures to incorporate it and even use it to its advantage was most sucessful if we judge from the self-congratulatory and arrogant tone of the decision of the 10th Session of the Central Committee(30-31 July 1983). The organization had subordinated so much of its vitality to the leadership without major disturbances, that the Central Committee could rubberstamp the process of its incorporation into the government's logic and functioning. Thus, the Movement's previously vibrant organization, which, in spite of its shortcomings and immaturity, was to be the basis, at least on paper, "of the socialist transformation of Greek society", had come full circle; it had, in effect, to submit its will and initiatives to the Government. Now only twenty-one months after the Movement's victory, "the Local and Professional (Clubs) as well as the Prefectural and Regional Committees are (were) responsible and accountable to the Central Committee, the Executive Bureau and the central guiding committees".[38] In this way, for the first time, the organization in effect recognised an organizational process which was opposite to the founding premises of a democratic mass Movement in which the rank and file was to exercise effective control over the leadership. The leadership was now to control the rank and file, making the organization an appendix to the government or, to put it cynically, to the prime minister since the Central Committee was meeting as a body only once a year, and the Executive Bureau was appointed by Papandreou himself. In other words, the decision of the Central Committee legitimized what was for the longest time the Movement's practice, and what we would call "the political proletarianization" of its membership. The organized base of PASOK was in effect seen in the same way that capital sees labour i.e. as a serving mechanism, a mindless organization which was to have no political will of its own or input to the further removed and neutralized governmental policies, which appeared by definition to be infallible.

Indeed, the decision of the 10th Session of the Central Committee was provocatively explicit about the new tasks of the Movement's membership. Thus, "the programme of activity" for the organization was "to inform the People...(and) support Governmental achievements". More specifically "the members of PASOK in their (day to day) work had to become the example for everyone to follow". That meant "the launching of a campaign at all levels for increasing productivity...(for) controlling the quality of production...(for) the creation of consumer cooperatives (which in turn) will promote the sale of Greek products...(and finally) for the materialization of the 1983 developmental programme".[39]

It was clear that PASOK's leadership, now occupying state power, was to remain faithful to its practice in opposition, keeping its own organization at the sidelines of political decision-making, and further consolidating its status as a mere observer of the political spectacle and mobilizer of support for its policies which were beyond question. In fact all the other tasks assigned to the membership are indicative of the government's effort to use the organization as a legitimizing factor for its policies. Thus the Central Committee's decision not only showed no concern about the nature of the mass entrance into the organization and in fact welcomed the development, but also set the ambitious task for the organization of further expansion by twenty-five per cent (from 200,000 to 250,000) before the Movement's first Congress which was promised for the beginning of 1984. Of course once more there was no reason given except that "the *allage* needs hundreds of thousands of activists to daily do battle from PASOK's ranks...(and) materialize the...political functions of the Movement",[40] of which at the time there were none, other than the governmental policies.

The self congratulatory tone of the 10th Session of the Central Committee should not be taken to mean that it met with no opposition. Two members of the Committee (N. Kargopoulos, M. Charalambides) while supporting the political proposal of the president, refused to align themselves with the other members and accept his organizational proposal. In fact Kargopoulos, who had recently resigned as secretary of the Organizational Bureau of the Movement, put forward a counterproposal to that of the president. From this counterproposal, which was distributed but never presented, we can gain a much more realistic picture of the situation within the organization and the overall policies and practices of the government of PASOK Kargopoulos' proposal was in fact "illegally" leaked to the press[41] and eventually led to the ostracization of its author from the Movement a few weeks later. It warned the Movement that "*allage* is moving backwards" as "the government is following the path of incorporation in the system...(a path) without return, and one followed by the 'socialist' governments of Western Europe". It was the first, and up to that point the only comprehensive criticism of PASOK's government and the Movement's overall organizational situation, in the post 1981 period. For that reason it is worth referring to even briefly, as it highlights both the post 1981 organizational situation of the Movement and the concerns of the disappointed membership.

After identifying, or in some cases predicting with admirable precision, the governmental policies to which we have already referred as unacceptable deviations from the Movement's programme, the document criticised the overall attitude of the government and the established relations between the latter and the party's rank and file. "The arrogance of power", Kargopoulos argued, "snobbism, lack of respect for the citizen and the people, are the main characteristics of this type of governing... favouritism has taken serious dimensions...(and) instead of a lever (of *allage*) the mass movement has become the enemy".[42] Furthermore, the document considers "the separation between the party's rank and file and the 'enlightened' governmental establishment arbitrary and authoritarian...(which in effect) pushes the Movement and the people into the corner". Finally, the "counter-proposal" suggests that this "course should stop" and that "in terms of timing, the Central Committee's session is the last chance for a radical change at all levels...(if we were) to convince the people".[43]

Needless to say, since the Movement was lacking any legitimate channels of dialogue and zymosis, the "counter-proposal" fell into a vacuum. Kargopoulos was isolated and soon after dismissed by the Disciplinary Committee of the organization.[44] Moreover, the proposal objectively could not have any further effect as it was full of contradictions. The third world dependency perception and tone which penetrate the document and which were very close to the C.P.'s language and practice (at least so far foreign policy is concerned), which in turn in the context of Greek politics contradicted the idea of the "democratic road", i.e. "democratic and pluralistic participation of the party of *allage*" which the counter-proposal supported passionately. The merit of this proposal lies in its descriptive value as it indirectly conveys the concerns of some of the Movement's membership and represents the first post-1981 organizational turbulence within the Movement, which like Cassandra was anticipating its future development.

The other important dimension of the Kargopoulos document is that it focuses on the ethics of the Government's practices and not simply on its politics. The document strongly questions the morality of the Government and the Movement's leadership "(since) slandering, petty conspiracies, petty calculations, provocations and immoral manipulations were on the daily agenda" of the organization. In fact this dimension of the Movement once in power, was beyond any doubt in the minds even of its most uncritical supporters since this immorality penetrated all its aspects and ranks from the bottom up. So shocking was the government-party practice that even the pro-government press graphically named its main instigators "godfathers of A*llage*".[45] The forced (temporary) resignation (6 June 1983) of one of the most prominent members of the Movement (A. Christodoulides, member at the time of the Executive Bureau and director of the Athens News Agency), who had been implicated publicly as gay by the right wing press was one of the highlights of this new socialist morality.[46]

The decisions and resolutions of the 10th Session of the Central Committee were unanimously adopted by the organization without problems or objections, setting pace in this way for the upcoming First Congress of the Movement which was to crystallize once and for all its organizational development. The Congress was to certify the death of PASOK as a Movement and its establishment as a "conservative" governmental party whose main concern was mobilization of support for state policies.

The First Congress: The Degeneration of the Promising Socialist Movement Completed

There are not many political parties claiming socialist titles for their organizations, which develop successfully under liberal democracy for over nine years without a Congress. Thus on that front (as in others) PASOK can claim to be internationally unique. As we have seen, the promise of the Movement's Congress functioned for the longest time as a fetishized commodity for the ideological and organizational unity of the party. But it was now, after having cleansed its ranks of possible threat to the leadership's will and having captured power, that the Movement could afford to fulfill its long standing promise and organize its First Congress. A major event like that could also reactivate the organization of the Movement which had been hibernating since the 1981 victory. This was needed not only because the Government was allegedly losing its popular momentum, but because the (European) election was approaching. The Congress was initially scheduled for the winter of the 1983. However, the Movement did not managel to put together its First Congress until May of the following year. It was an event which not only rubberstamped the organizational development of the Movement up to that point, but its preparation, its actual organization and its celebrated three day session highlighted in a most vivid fashion PASOK's major characteristics both as a party structure and as government.

The Congress's entire preparatory period highlighted the sloppiness and the superficiality which had characterized the Movement's development up to that point. First of all, the pre-

The Movement in Power

paratory committee postponed the date of the Congress twice before finally scheduling it for 10-13 May 1984. The political proposal of the Executive Bureau was published only five weeks before that date, leaving very little time for the rank and file to undertake any meaningful discussion or input. The selection of the delegates of the alleged 220,000 member organization once again was neither a smooth nor a democratic process. During this preparatory period the direct intervention of the leadership in the selection of delegates continued. Many members lost, or found their membership cancelled, while the familiar phenomenon of manipulation of the entire existence of well established clubs of the organization was not rare (The latter induced merging of local clubs, mass participation of activists of the central offices in the meetings of the local clubs with tendencies towards criticisms of the Movement etc).[47]

Furthermore, the rank and file appeared to be left in the dark as to the technical and logistic aspects as well as to the overall purpose of the Congress. For example, while just six weeks before the Congress, the guidelines of the preparatory committee designated the number of delegates at around 2000,[48] the actual number was under mysterious conditions inflated to 3000.[49] Obviously this multitude of delegates not only guaranteed limited intervention from the membership on the floor of the Congress, but it also allowed the leadership to give delegate passes to its favoured government functionaries who had no relation to the party's actual structure.[50]

The political and the organizational tasks of the Congress were not any clearer. Of course there was the long standing promise to finally vote on the Movement's constitution, but the immediate tasks of the Congress regarding this project, such as the election of a new Central Committee, were rather vague and confusing, as improvisation and the "play it by ear" mentality of the Movement were once more at their height. Thus while the preparatory committee claimed the size of the new Central Committee to be 120, and on the eve of the election, the government's spokesman said that this number would be limited to 100, the new Central Committee emerged from the Congress numbering 140.

The actual procedure of the three day Congress was not any better. The President delivered an overpowering three and a half hour speech within closed doors leaving only thirty seconds for each delegate's intervention. In fact, the 1977-1981 report of action of the Central Committee was rejected only by sixteen delegates while ten others abstained, out of a total of 2381 delegates who voted on the matter.[51] PASOK's peculiar conception of democracy was once again in action. Of course there was occasionally a superficially critical tone in some delegates' comments from the floor of the Congress on issues such as the "inertia of the party's organization", and "the phenomena of the disfunctioning" of the governmental apparatus as a result of the Movement's intervention. However, since the procedure strictly prohibited actual dialogue, these criticisms served only the personal ambitions of individual members indirectly legitimizing the overall undemocratic procedure. The criticisms of Papandreou's son George, which could not help but remind us of Andreas' own disputes with his father twenty years earlier and which contributed to building the young Papandreou's radical public image and popularity, were a prime example of the nature of the "opposition" voiced at the Movement's first Congress.[52]

The proposals introduced and the actual decisions of the Congress did not fail to highlight the Movement's most controversial aspects. To begin with there was a significant difference between the proposal of the Executive Bureau distributed to the organization ahead of time, and the tone of Papandreou's address to the delegates. Thus, while the proposal of the Executive continued the self-congratulatory and arrogant tone of the Government, Papandreou's proposal appeared more subdued, with strong self-critical overtones about the achievements of the Government and the practice of the organization of the Movement. In fact Papandreou went so far as to accept failures in some aspects of the Governmental policies and to admit to some weaknesses in the actual functioning of the Government deriving from the wrong attitudes of the new state personnel and the party's membership. Contrary to merely condoning the development of Governmental policies and practices up to that point, in his speech Papandreou went so far as to admit "the lack of coherence in the Governmental measures, the delays in the synchronization of our (Government's) priorities and the implementation of intermediary goals".[53] Furthermore he identified the "degenerating phenomena of the arrogance" of various parts of the Government as the primary reason for leaving the people uninformed, which in turn "had prohibited the development of real and decisive dialogue (between the people and the

The Movement in Power

Government) before the institution of the basic reforms".[54]

There was little doubt that Papandreou's radical tone, by virtue of his self-awareness, aimed at diluting the growing disappointment and discontent not so much within the Movement's rank and file but mainly among the Movement's supporters at large. It was a task that only he, as the undisputed genuinely popular leader with magnetizing charisma could do without alienating the Movement's activists and the Governmental functionaries, who had instigated and benefited from the practices which he criticized. Papandreou's dominant and domineering position both in the Government and in the party meant that the latter's political survival depended exclusively upon him. The enthusiastic response to the speech, its unanimous adoption by the delegates as the framework of the political resolution of the Congress and finally the silence of the Executive, whose proposal had just been severely undermined, are strong cases in point.

Furthermore, the politics of Papandreou's proposal, which differed little from the proposal of the Central Committee and which finally became the political resolution of the Congress, maintained in an admiring fashion, the flavour of the Movement's politics which were full of contradictions as they tried to wed the old radicalism of the Movement necessary for the reactivation of the organization, with the cynical realism of the Government. Thus, while the Congress once again confirmed its faith in the "3rd of September Declaration", in the Movement's analysis of the Greek society along the lines of dependency theory and even reaffirmed the "acceptance of Marxism as a method of social analysis",[55] its social and economic commitments were in striking discord with its analytical premises. Renewing its commitment to "a Peaceful, Democratic and Greek road to Socialism" and briefly to the "Third Road,"[56] the Decision of the Congress indicated that this strategic goal was linked with the development of a "mixed economy both at the level of structure and at the level of relations of production".[57] What this meant was the development of the Greek economy in three complementary sectors: "public, private and social experimentation",[58] which in turn were somehow to guarantee "the new structure of the relations of production".[59]

Of course, faithful to the Movement's tradition, the Declaration of the Congress maintained the radical rhetoric on issues such as the country's foreign policy, its relations with NATO and the EEC, as well as its concerns on world peace, and on the environment, quality of life, and issues related to the position of women in Greek society. It was a rather peculiar expression of the Movement's old radical flavour as its record in Government was anything but favourable or responsive in any serious way to these issues. In fact, as we have seen, some of the Government's policies were in direct conflict with them. But again this was not suprising since the Movement in its more than nine year history had displayed an incredible capacity to claim one thing while doing another, and to convince almost everybody that this was consistent with its original commitments. This was an ability based on the one hand, on the personality and the charisma of Papandreou,[60] and on the other on the low political education of the population, or what we have called its "a-political over-politization" to which the parties of the opposition also contributed.

A New Role for the Rank and File

Despite the apparent difference in the political tone of the Executive's proposal and Papandreou's own speech at the Conference, the leadership of the Movement appeared united on the organizational question of the party. In Papandreou's own words "the oppositional dynamism of the organization inhibited its necessary reorientation when PASOK found itself in government"[61] and therefore organizational adjustments were becoming urgent. To put this another way, the party/Government leadership wanted to restrict the organization from becoming even more of an obstacle and embarrassment to the Government which was cruising backwards, away from its pre-electoral promises, and in many cases, as we have seen, contrary to its overall socialist premises. Thus, the new Constitution attempted to establish "the proper dialectical relationship between government-party and party-State".[62]

What was really hidden behind this presentation of the leadership's intentions was the need for a new role for the organization. The organization was to become a mere appendage to the government's legitimizing function. The new role preserved for the Movement's membership was "...to support, analyze and adopt the accomplished governmental work". According to the new constitutional principles "the organization of PASOK had to connect the past with the present and the future, the general with the particular...(which) in the context of today's reality (meant) the support of the governmental policies of the possible".[63] Defining the organization's functions along these lines, Papandreou and his inner-cicle hoped with the continuation of membership co-optation through recruitment into the Government, to both tame the organization's opposition and further to use it as a mobilizer of support for the Governmental policies which were becoming less and less popular as the Government was turning towards the economic initiatives of austerity. Moreover it is worth noting that this explicit effort to redefine the relations between the government and the party coincided with the Government's restructuring of its own apparatus. The latter became necessary due to the increasing attempts of the Government to further centralize the governmental and state structures.[64] Thus, for obvious reasons the new conception of governmental functioning needed the least possible party interference as well as a further mobilization of support. This was exactly what the leadership was hoping to achieve through the new Constitution of the Movement, underlining once again, in the most vivid way possible the dialectical relationship between the party and the Governmental agencies.

Of course the Movement's decision on its relation to the Government was not so crude as it might appear from the above exerpts. As we have seen, PASOK has made the habit of introducing controversial actions dressed in the most deceiving rhetoric. Its Constitution was not about to be the exception. The new restrictive framework for the functioning of the organization was accompanied by a number of premises in striking contradiction to the effort to put the party structure on the back burner of the Movement's political agenda and to rely more and more on government-state expertise. Thus, in the midst of its effort to limit the organization's impact and interference in Government policies, the organizational resolution of the Congress advised the membership to continue its involvement in the mass movements. It reaffirmed that "democratic dialogue is the prerequisite for the (further) politicization of the organization". And further, it argued that the support of the organization for the Governmental policies which "guarantee the march to socialism", was not necessarily a one way street, and it expressed the hope that the organization "would be in a permanent state of peaceful revolution so that it would not become bureaucratized".[65]

However, this built-in dimension of the new constitution was not enough to dilute the whole process of the neutralization of the organization by incorporating it into the Government logic. This is in fact what was argued in the context of the general organizational priciples of the Movement, which acted in turn as a natural barrier to the possibility of pushing apparently democratic rhetoric to its logical conclusion. These organizational principles were captured in the twofold slogan of "Democracy and Efficiency" which had been used since the pre-1977 period. As it was understood,[66] this fundamental organizational principle meant nothing more than the maximization of the effectiveness of the organization's further efforts to expand its recruitment and materialize the idea of intra-organization unidimensionalism. This is also what derives from the condemnation of the organizational practices of autonomy (between the party and the mass movements) and internal pluralism,[67] which were previously praised as the Movement's advantageous point of differentiation from the traditional left. Thus, in effect, by taking the "movement" out of the Movement, the new organizational principles constitutionalized the direction of PASOK's development until that point. It was a direction which turned toward the creation of a unilateral and "conservative" party.

The Congress and the new Constitution, concretized in a firm and official fashion the organizational practices of the Movement. In addition to the issues related immediately to the taming of the organization, the Congress crystallized all the hierarchical, undemocratic and centralized tendencies which PASOK's internal opposition had attempted to fight as the Movement was struggling to develop its organizational and political identity. Thus the "new" organization which arose from the Congress presented no surprises, except that now the proscribed organizational patterns were not "common law" types of practices, but rather were written principles explicitly approved and legitimized by the higher collective body of the Movement – the Congress. The Constitution is full of clauses which, on the one hand, guarantee the perpetuation of the centralized tendency of the organization in which the "President and the Executive Bureau" have the last word on all major issues, while on the other, they establish numerous disciplinary measures for the rank and file who dare to step outside their duties. Indicative of the new organizational document of the Movement is the list of the membership's duties which is strikingly longer than that of their rights (i.e. ten vs four), while thirteen out of the total eighty-eight articles of the Constitution deal directly with the disciplinary functions of the organization.[68]

However, what best captures the character of the Congress as a mere rubber stamping of the Movement's organizational development to that point is the explicit as well as tacit recognition of the role of A. Papandreou in the party structure. The chorus of events which secured Papandreou even further as the omnipotent source of power in the Movement, began with the silent and obedient way in which the members of the Executive Bureau in effect accepted the denial of their political proposal in the President's speech. In addition to this significant but symbolic event, the Constitution preserved an explicitly exceptional role for the President when he was given unlimited powers (e.g. full representation of the Movement and capacity to decide on any issue if the circumstances do not allow a session of the Executive Bureau, which in any case is called in session only by him – article 62). In addition, he also remained outside the collectivity of all elected bodies of the organization and was therefore accountable to no one. Thus the First Congress of the Movement, held almost a decade after the Movement's creation, not only did not challenge his position but did not even bother to put on a show of acclamation for the most important position in the organization. The controversial equation, PASOK = Papandreou, established by the practice of the Movement up to that point was thus not only tacitly recognized but it was also constitutionalized.

The First Congress of the Movement had been seen as the ultimate chance of the rank and file to be heard, and had served for the longest time as a fetishized organizational commodity keeping within the organization those members who wanted to have a greater say in the Movement. However, it became in effect merely a public relations extravaganza, as the well-organized publicity and glamour gained by the participation of radical organizations from all over the world in the opening ceremony indicated. It was also the point at which the arbitrary and controversial organizational development of PASOK was officially and legitimately approved. The dialogue before and during the Congress, was, as we have seen, severely restricted; politics were not discussed in any meaningful way, and even the election of the Central and Disciplinary Committees of the party, which ended up being the highlight of the entire procedure was marked by the President's dominance and the egocentric petty interests of the delegates. As we have mentioned, the Congress decided in a mysterious way to increase the size of its Central Committee from the originally planned 120, to 140. Obviously the anonymity of this multitude would further reduce the liklihood of the development of a counter-pole of power in the Movement. Papandreou himself nominated 95, out of which 88 were elected. Interestingly enough, all nine members of the Disciplinary Committee were in fact his recommendations.[69] In addition there is strong evidence that even this strange version of intra-party democracy was not a smooth process as many candidates tried to manipulate the result.[70]

An Assessment

The organization, the preparation, the actual proceedings and the outcome of the delayed First Congress of PASOK proved in a crystal clear fashion that the Movement's collective procedures, organizational principles, Constitution, and elected collective bodies (including the Central Committee) were not really important to its development. If the Congress accomplished anything, it was to crystallize the organizational practice of the Movement up to that point and to consolidate once and for all the single source of power in the party – Andreas Papandreou. The Movement's structure was downgraded and pushed permanently into the corner as it was incapable of having any real impact on the party's politics. This of course does not mean that it was killed off. It was in fact kept sufficiently alive to function as the enthusiastic audience of the leadership's initiatives, as a mobilizer of support for the Government's controversial policies and as an organizational appendage necessary to its electoral successes, something which was proven vividly in the 1984 and 1985 elections. Its internal functions, organizational principles (of democratic procedures, of mass participation of sexual equality etc.). although severely circumvented, were to function within this prudent framework, which had a twofold purpose. On the one hand it was to work as a fetishized activity to keep the organization alive enough to perform all these

The Movement in Power

legitimizing functions, and on the other to act as a smokescreen for the arbitrary and undemocratic initiatives of the Government.

Through the Congress, the Movement's leadership had not only dealt efficiently with its internal problem of hyperactivity and rid itself of this burden but it had also managed to make the organization into a factor which would be conducive and legitimizing to its political choices. From then on Papandreou and his close associates would have no problem in claiming one thing and doing another,[71] and without the internal turbulence capable of reversing that pattern. The fact that even the 1986 "mass" resignations of trade unionists did not cause any long term problems for the Movement while at the same time it managed to mobilize enough support to legitimize its controversial initiatives[72] is a case in point. The organizational principles and the Constitution were transformed into malleable formalistic devices which could be shaped at the absolute discretion of the leadership according to the political needs of the circumstances. The amendment of the Constitution in the 14th Session of the Central Committee just two months after the Congress is a clear example of this pattern.[73]

Of course the outcome and the implications of the Congress did not go unnoticed by some members of the Movement. The expectations vested in the organization's First Congress during the Movement's nearly ten year development and its use as a device of manipulation for the legitimation of the most arbitrary and undemocratic organizational initiatives of the party could not be easily forgotten by its keenest members. Thus, it was of no surprise that in the aftermath of the Congress there was a diffuse disappointment and reaction to the actual accomplishments of the Congress. However, this reaction was bound to be harmless and incapable provoking even the slightest turbulence in the party's structure. The few pre-existing outlets for dialogue and collective reaction to the leadership's initiatives, which had in the past assisted the development of internal opposition, were permanently sealed off. Thus the only response left on the part of the membership which stubbornly insisted on thinking about and questioning the Movement's direction was that of departure. In fact there was a new and silent exodus from the party in the months following the Congress.[74] This time it appeared impossible to create even the mildest waves within the organization.

The Movement in Power

The Congress consolidated the degeneration of PASOK into a monolithic organization. Yet it was based not on a tight organizational and manachean type of politics but rather on the chaotic polyphony of the membership. It is a polyphonic pluralism which is very much entrenched in the Governmental policies. These include the simultaneous promotion of business spirit and of progressive political art by the ministries of Youth and Culture; support and participation in the peace movement while at the same time militarism is kept alive more than ever before; and the coexistence of solidarity for the class struggle with disciplinary initiatives against the working class. However, the contradictions are only apparent. They are in fact the very essence of PASOK's existence and strength. Everyone can believe and argue politically on just about everything (e.g. from claims on the heritage of Marxism-Leninism and the concepts of self-management to the most liberal belief on welfarism) so long as the last word on the subject is left to the only source of the Movement's power (i.e. Andreas Papandreou) whose omnipotence is not to be challenged. PASOK lacks those channels of dialogue in which the resolution of political and theoretical conflict is supposed to take place. Consequently the leadership can use the party's chaotic internal situation both as the framework of its political flexibility and as an alibi for its undemocratic practices. This was the organizational result of the death of a genuinely democratic party capable of structurally transforming Greek society.

The deterioration of PASOK's organization which had begun with the split of 1975 and reached its height during the events of 1977, was completed, ironically enough, at the first national major collective function of the Movement – the Congress. In a most definite and strict fashion the Congress had framed the future development of the first mass socialist party of Greece. After the 1984 Congress, all of PASOK's organizational and political developments were to be merely permutations of the established organizational and political patterns. The experience since the Congress constitutes the depressing and living truth of this conclusion. Even the much advertised Twenty-second Session of the party's Central Committee (31 May 1987), which called for the *"renaissance"* and the reconstruction of the Movement's organization appears not to have departed from the established pattern of PASOK's development. This is a pattern, which neither the upcoming Second

Congress of the Party nor the intention to "reconstruct" the organization to "meet the challenge of the year 2000"[75] afford any room for change in such a personified and centralized organization. This becomes even more obvious when we take the state of affairs within the opposition into consideration.

The Opposition

The story of PASOK as a government and as a party constitutes a long chain of broken promises, undemocratic practices and dashed hopes. It is therefore only natural to expect the growth of opposition and the creation of alternative political formations. However, this is far from being the case. Much of this is to be credited to Papandreou's capacity to neutralise his internal and external critics. But it must also be attributed, if not to a greater degree, to the capacities or rather the lack thereof, of the prospective opposition.

The 1981 electoral result was as much a defeat of the right ("New Democracy") as it was a victory for PASOK. The June 1985 election was nothing but a repetition of the 1981 situation. After finding itself for the first time outside of state power, "New Democracy" has experienced great difficulty in appearing in any united fashion. In reality, the dominant right wing party is deeply divided between a faction which favours modernization along Keynesian lines and political development along the lines of "modern liberal democracy", and another which is clearly more conservative and maintains a discourse reminiscent of pre-1967 clientelistic practices. It is a deeply rooted division to which the frequent changing leadership of the party has been addressing itself, not without major organizational problems.[76] Thus it took some time before New Democracy was able to overcome its problems and present a threat to the Socialist omnipotence, as for example when in the municipal elections in the fall of 1986, the Right won the mayorship in the three major cities of the country. Up until then, New Democracy's opposition against PASOK had been inconsistent, scattered and very much dependent upon the character and the mood of its leader at the time (New Democracy has changed leadership three times since 1982). It is only in the last couple of years that New Democracy under the controversial leadership of K. Mitsotakis has managed to perform some serious and effective opposition (judging that is from the Right's successes in the unions, associations and the student movement). And it is this development which has resulted in the revitalization of right-wing hegemony of late. The major obstacle in New Democracy's path towards a return to power appears to be Papandreou's skillfull attack on its most conservative elements and his effective use of the 1981 atmosphere of polarization between the "democratic forces of progress" (of whom PASOK is the representative), and the conservative and regressive Right.

The weakness of the Right-wing opposition had been exceeded by the inarticulate practices of the left. The orthodox Communist Party, the most significant left wing rival of the Government, has not managed to put forward a comprehensive opposition to PASOK. In spite of its organizational militancy and effectiveness, its opposition has been limited and scattered. It usually focuses on the day to day issues with a special emphasis on the Government's foreign policy – especially the part referring to the country's relationship to NATO and to the US. On other issues, the C.P., a victim of its own progressivist assumptions, usually presents a rather schizophrenic image, since it simultaneously welcomes some of the Government's policies and critizes some of the others for "not going far enough".

Furthermore, the C.P. can be, and has been, easily manipulated by the skillfull leadership of PASOK whenever it has attempted to step outside the boundaries of "constructive opposition" and beef up its militancy. Trapped in its self-imposed obligation to support Soviet diplomacy, and being itself a victim of instrumentalism, the C.P. has put itself on a self-neutralizing course. Phenomena such as abandoning oppositional activity when the Government makes a positive gesture to one of the Soviet bloc countries or when its representatives are included in the corporatist initiatives of the Government, are not rare. During PASOK's first term, many critics of the party – both on the left and the right – talked about their suspicions of a secret "moratorium" of peace between the C.P. and the Government. The truth or falsity of this speculation is rather irrelevant since it is the internal/structural limitations of the party which themselves make it incapable of confronting a left wing government in an imaginative way.[77] These problems of the C.P. have been met with unusual (for that party) internal criticism from its membership, especially after its electoral losses in the last election. Of course this internal criticism has been confronted with well known Stalinist methods. Growing numbers of rank and file, as well as middle level functionaries have been kicked out of, or forced to resign from, the party.

Finally, the C.P.'s stifling organizational structure cannot present a positive alternative to PASOK's disappointed membership. The Party has not only largely ignored PASOK's internal developments but also since it has itself basically the same undemocratic internal practices (although nearly not so personalized as PASOK's), it cannot be a realistic point of attraction to the Movement's sceptical activists. PASOK's initial promise for "democratic procedure", its rhetoric and the facade of internal democracy performed through its chaotic pluralism has created at least an illusion among its membership that they belong to a democratic party, which they were not ready to leave for a party which had little or no consideration for democracy. All these problems of the C.P.'s opposition to PASOK did not change even after the former's Twenty-second Congress, where it appeared to have adopted tougher stands *vis-a-vis* the Government. Its main new call for a strategy for *"Allage* towards Socialism" is neither concrete nor convincing.

On the other hand, the record of the C.P./Interior in opposition has not been any more impressive. This party confonted the Government's policies with the strategy of "critical support" (especially during the first year of Socialist rule when the party was overwhelmed by PASOK's popularity). The result was a lack of focus in its policies. It usually flip-flopped between enthusiastic support for the Government's reforms and "disappointment" in PASOK's broken promises. After 1982, with the change in the Government's policies and given its dismal showing at the polls, "the Interior" launched an attack on the Government. This attack however, did not develop into the formation of comprehensive policies – only the jargon (often workerist) and the tone have changed, and they are not particularly convincing. Furthermore, its seventeen year history of reformist policies and lack of an industrial working class base, along with its limited resources and its miniscule and periodic parliamentary presence (the party's parliamentary representation was limited to the European Parliament while it had no members in the 1981 national Parliament[78] have set the limits of its support and effectiveness. In addition, the moderate political tone of this party which derives from its strong Eurocommunist tendencies, could not have been a political alternative to PASOK's radicalized rank and file, which has not been trained to make the fine political distinctions upon which this party's politics are based. Thus, even during the periods of intense scepticism and wide-spread pessimism among PASOK's membership, the latter appears to prefer to stay in the Movement, which at least on the surface has kept its radical rhetoric.

What made this party interesting however, were its differences from the (orthodox) C.P. Apart from not being slavishly pro-Soviet, the C.P. of the Interior also embraced (albeit somewhat opportunistically) the political concerns of the social movements. With the exception of its loose links with the Eurocommunist parties, the C.P./Interior did not have any restrictive internationalist commitments and its structure was, at least in theory, more democratic. For these reasons it became by default, the point of political reference for the independent left, which was concentrated around numerous theoretical publications. In fact the existence of these groupings had been recognized as the main reason for its survival. Confronted with its shortcomings, the party in fact opened its organization to them in an effort to develop an alternative strategy for the left. A new party was in fact created – the Greek Left (Ellenike Aristera) in April 1987, which is hoping to become a truly radical and effective alternative to PASOK and its controversial reforms.

In sum, the situation in the ranks of the Movement's external and internal opposition, in the first five years of Socialist rule did (could) not create any problems for PASOK's reign. The admirable manoeuvrability of its leadership, the disgust in collective popular memory for right wing rule and most importantly the state of affairs in the parties of the opposition guaranteed the absense of any serious challenge to the Socialist government. But since 1985-1986, this situation has changed. The forced ostracization of the left tendency which was integral to PASOK's strength, the revitalization and reorganization of New Democracy in combination with the fading away of the memories of the forty year long right wing rule make PASOK's future anything but secure. Moreover, there is a major question in PASOK's future as to how long the people of Greece (especially the radical urban working class and new middle class of the disproportionally large service sector) will put up with the austerity programme and/or continue to be tricked by the Socialist controversy. The latest scattered but militant expressions of these strata point to the urgency of the need for a new genuinely democratic socialist political representation. If this does not happen – and the ball is objectively in the court of the left – the danger of awarding the frustrations of these strata as a gift to the right is increasingly becoming a real possibility.

Notes

1. *The Programmatic Declarations of the Government*, Athens, General Secretariat of Press and Information, 1981.

2. *The Greek Economy in Figures: 1984.* Athens: Electra Press Publications, October 1984.

3. K. Zephyros. "E Ekonomike Politike tou PASOK yia to 1983" (The Economic Policy of PASOK for 1983). *Marxistiko Deltio* (Marxist Bulletin). No. 17, January to March 1983. p. 16.

4. *Agora*, Vol. 2. 22 June 1987. p. 42 and P. Linardou – Rilmon. "Istorike e Ptose Paragoyes, Paragikotetas kia Ependyseon ste Viomechania Kata to 1982" (The Historic Drop in Productivity, Production and Investments in Industry During 1982). *Oikonomikos Tachidromos* (Economic Postman). 16 June 1983.

5. The figure is only indicative as in addition to the shortcomings of the technocratic measurement of unemployment and the traditional Greek aversion for systematic statistical research and analysis, there is disproportional underemployment in the country due to the swollen petty commodity production and the seasonal nature of tourism which occupies a dominant place in the Greek political economy. There is little doubt however, that unemployment was reaching unprecedented heights especially in the beginning of PASOK's second year in power. P. Linardou – Rilmon. "Entone Afxise tes Anergias Sto A' Trimeno tou 1983" (Intense Increase in Unemployment in the First Three Months of 1983). *Oikonomikos Tachidromos* (Economic Postman). 16 June 1983.

6. OECD *Report on Greece*, January 1986. Quoted in J. Petras, "The Contradictions of Greek Socialism", *New Left Review* No. 163, May-June 1987.

7. Ibid.

8. Ibid.

9. Ibid.

10. Many obervers have attributed PASOK's remarkable re-election in 1985 to its decision to withdraw its parliamentary support from K. Karamanlis for the renewal of his term as the president of the republic.

11. Well before Papandreou's opening statement in the new parlia-

ment, where the promise of the continuation of PASOK's programme in a "realistic fashion" was given, we witnessed arguments for the austerity measures in the speeches and announcements of K. Simites, then new minister of National Economy and D. Chalikias, director of the Bank of Greece. See *To Vema.* 23 June 1985 and *Ta Nea.* 12 August 1985.

12. In his speech to a conference of European and American economists, A. Papandreou said about the problem of unemployment in Greece: "Of course the figure 8% is a conservative one. I believe that 9% or even 10% is closer to reality". Summer 1983.

13. P. Linardou – Rilmon. "Istorike e Ptose Paragoyes, Paragikotetas kia Ependyseon ste Viomechania Kata to 1982" (Historically Significant the Drop in Productivity, Production and Investments in Industry During 1982). *Oikonomikos Tachidromos* (Economic Postman). 16 June 1983.

14. It is worth noting that this Socialist "achievement" was applauded by such international organizations as the IMF and the OECD in their 1984 reports.

15. Many observers have claimed that the agreement resembles the one signed by the Philippines. Basically the agreement guarantees the presence of the US bases for five years and then further negotiations are anticipated.

16. As a result of some of these policies, Western leftists have often made the mistake of exempting PASOK from the criticism applied to other Euopean social democratic governments. For example see: J. Petras "The Rise and Decline of Southern European Socialism" *New Left Review* No. 146.

17. See Chapter I, also M. Serafetinidis et.al. *op. cit.* and V. Choraphas, "Shipping Capital in Greece" *Antitheses.* No. 15. (in Greek).

18. For a critical presentation of PASOK's record on that front, see Eleni Stamiris, "The Women's Movement in Greece", *New Left Review,* No. 158, July-August 1986, pp. 104-112.

19. This appears to be particularly the case after the post 1985 election's heated jargon on the need for the implementation of the "Stabilization Programme". See A. Papandreou's speech to Parliament in inaugurating his second term in power, *To Vema,* 23rd of June, 1985.

20. To be more specific, as we saw in chapter I, the base of the participation of Greek capital, even until World War II, in the international division of labour was its role as intermediary in the Balkans and the Middle East. At the same time, Greece's internal development was rather modest. It depended heavily on state support and was based primarily on agriculture, light industry and the development of the country's infrastructure. After the war however, the Greek economy's ties with the world market were primarily its shipping capital, tourism

and of course its emigrant labour. In comparison, the development of the secondary sector remained modest. This pattern of development worked relatively smoothly up until the beginning of the 1960s when in nascent forms it started to display its shortcomings and signs of its difficulty in reproduction. These difficulties did not become obvious until 1973-1974, when the world capitalist crisis and the beginnings of the new international division of labour put this pattern of development to the test. Shipping was confronted with more and more pressures, tourism started to show signs of saturation and emmigration ceased to perform in its role of safety valve. The changing nature of the Middle Eastern societies as centers of major capitalist accumulation indicated that Greek capital could only survive and expand if it was to adapt to the new circumstances. It had to reorient itself along productive and industrial lines and towards an export oriented production. In addition, it was recognised that Greece could become, especially after the destruction of Beruit, the transportation bridge and the economic capital of the region. But is was not only these considerations which made the argument about productivity powerful, but some other internal factors which were even more pressing. These were the increasing incapability of domestic production to respond to the needs of the economy, which in turn, as we saw, had widened the gap between imports and exports, increased the foreign deficit, augmented the overall public debt, and seriously crippled foreign reserves, which were putting constant pressure on the drachma.

21. Quoted in Karabelias. op. cit. p. 52.
22. Greek dailies. 15 May 1982.
23. G. Arsenes. "Greece's Rising Economy Enhances Investment Climate". *Greek Investment News* 1 No. 1, April-May 1985.
24. Many political observers make the point of linking the Prime Minister's withdrawal of the measure with the parade of unscheduled visits of some ministers and senior M.P.s to Papandreou's office. The failure of Papandreou to even consult the minister responsible for the legislation resulted in a mini cabinet crisis and the resignation of the minister M. Drettakis from the Government and the party.
25. After the 1985 election, for "efficiency reasons" the cabinet was reduced to only 19 members, of which one ministerial portfolio is held by Papandreou himself and three others are ministers "under (i.e. reporting directly to) the prime minister", is the latest example of this tendency.
26. This tends to add little credibility to Papandreou's controversial decision in the spring of 1985, to reform the constitution and remove Karamanlis (whose charisma and national appeal undoubtedly made him the other major pole of power in the Greek political configuration) from the Presidency and replace him with someone of Papandreou's own choice. This was designed to further promote and consolidate the

tendency toward one man rule.

27. For example, the crack-down on an independent anti-fascist political initiative in November 1984, the virtually forced imposition of partisanship in the labour, women's and students' movements, and the partisan criteria for subsidizing the artistic community – are only some cases in point.

28. That was at least the distinct impression of the writer when he attended one of these meetings (Piraeus 4 July 1983). The Minister, who was supposed to initiate the "dialogue" (G. Moraites), referred extensively to the Governmental achievements, turning the dialogue into a monologue. As well, all the critical questions from the audience were either removed arbitrarily from the floor by the chair, composed of the local party officials (maybe in their effort to show a good face to the central leadership) or were squeezed out by time constraints.

29. "Deka Chronia PASOK" ("Ten Years of PASOK") *To Vema.* 13 May 1984.

30. Four M.P.s of PASOK, two of whom held ministerial portfolios resigned from the party in its first year in power: Petsos (Deputy Minister of Defence), Hondrokoukes, Bouloukos and S. Panagoules (Deputy Minister of Internal Affairs).

31. From Papandreou's speech to a mass gathering (4000 functionaries). *Mesemvrene* 1 February 1982.

32. In addition to internal memos, PASOK organized rallies of its functionaries similar to the one described above in footnote 31. There again the spirit of Papandreou's speeches was the same. See *Ta Nea* 4 October 1982 and *To Vema* 28 November 1982.

33. The disciplinary interventions in the Prefectural Committees of Rhodes, Corfu, Prevezas, Argolis, and Karditsas, the prosecution of the Lesbos Committee by the Disciplinary Council of the Movement and finally the crude administrative and paternalistic intervention in the life of the organization in Larissa during PASOK's first term in power are nothing but cases in point.

34. The figure was given by Papandreou in his speech/proposal to the 10th session of the Central Committee of the organization, in *Enemerotiko Deltio* (Information Bulletin), special issue, September 1983.

35. Even if this figure is a little inflated, it remains a very impressive one. This is especially so when we consider the respective figure for PASOK's Spanish counterpart PSOE, whose membership in a country with more than three times the population of Greece does not exceed 150,000.

36. S. Kouloglou, *Sta Ichne tou Tritou Dromou* (In the Footsteps of the Third Road). Athens: Odysseas, 1986.

37. There are huge line-ups of people, who are members of the organization, during office hours of the Bureau every week at the Move-

ment's headquarters. Most of them are there for favours, the overwhelming majority of which are related to job search. Personal witness.

38. *Exormise*. 6-7 August 1983 and also "E Apophase tes 10es Synodou tes Kentrikes Epitropes tou PASOK" ("The decision of the 10th Session of the Central Committee of PASOK") *Enemerotiko Deltio op. cit.*

39. Ibid.

40. Ibid.

41. Parts of the twenty-nine page "counter-proposal" (*Mimeo*) were published on the first day of the 10th Session in *Elefterotypia*. 31 July 1983; also published in the journal *Scholiastes* September 1983.

42. Ibid.

43. Ibid.

44. After his departure from PASOK, Kargopoulos established the "Independent Socialist Party of Greece" (A.S.K.E.), which enjoys a marginal existence on the fringes of the Greek left.

45. M. Demitriou, "Pioi Alithina Synomotoun sto PASOK" ("Who Really Conspires in PASOK") *To Vema*. 1 July 1984.

46. Christodoulides had been involved in PASOK since the days of PAK and was one of the very few radicals who remained in the Movement's hierarchy. The rumours at the time of his ostracization from the organization had it that he was the victim of the orchestrated slandering of the *palaiokommatikoi* nomenclature of Papandreou's inner circle who wanted to further limit the influence of the radicals. Whatever the case might be, one is certain that neither the Executive nor Papandreou himself objected to his resignation as they remained completely silent at the time. Christodoulides was in fact the 14th member of the Executive Bureau that has resigned from the Movement in the organization's since 1974.

47. These phenomena did not take dimensions severely critical of the organization's unity, since to a great extent, for reasons we have already mentioned, they remained isolated. The unprecedentedly large membership, most of it new, and the absence of patterns of communication within the organization largely took care of this. However, activists who kept a close eye on the situation or with some personal connections with the Movement's rank and file had no trouble to identify this interference. Interview with D. Tzouvanos, 12 July 1984.

48. *To Vema*. 18 March 1984.

49. According to at least one reliable source, contrary to the organizing committee's plan, approximately 500 members became "ex officio" delegates on the eve of the Congress. D. Tzouvanos *op. cit.*

50. *Fylladio*. No. 9, May-July 1984. Special Edition on PASOK's Congress.

51. *To Vema*. 13 May 1984.

52. Young Papandreou's criticisms rested on a mere description of the phenomena of gradual isolation of the party from the people and underlined the need for a more open party structure. *To Vema.* 13 May 1984. However, he did not even try to find an explanation, which the reader of this book by now knows goes beyond the disfunctional symptoms of the post 1981 situation.

53. A. Papandreou, Speech to the First Congress of PASOK, *Ta Nea.* 11 May 1984.

54. Ibid.

55. "Oi Theses tes K.E. tou PASOK yia to Synedrio" ("The Theses of the C.C. for the Congress of PASOK") Athens, Publications Bureau KE.ME.DIA.-PASOK, 1984, pp. 213-214.

56. Ibid. p. 214.

57. Ibid. p. 114.

58. Ibid. pp. 114-118.

59. Ibid.

60. The fact that on numerous occasions Papanderou had to make special appearances on national television every time his Government was in some kind of public image trouble is a good case in point.

61. A. Papandreou, Speech to the First Congress, op. cit.

62. "Organizational Proposal" in the "The Theses of the C.C. for the Congress of PASOK" op. cit. p.187.

63. Ibid. p. 184.

64. *To Vema.* 11 March 1984.

65. Ibid. pp. 183-191.

66. Ibid. pp. 216-217.

67. Ibid. p. 184.

68. Ibid. pp. 218-240.

69. *Ta Nea.* 15 May 1984.

70. The unusually late announcement of the results of the election of the Central Committee, the number of appeals and finally other second hand evidence (D. Tzouvanos, interview op. cit.) have underlined a number of irregularities in the elections to the Central Committee.

71. Even the composition of the new Central Committee provides us with a good example of this habit. After having passed a resolution committing the Movement to the promotion of equal participation of women in the organization, the Congress did not manage to elect more than eleven women out of the total 140 to the Central Committe.

72. Calling on the Movement's trade union membership to support some of the Government's highly questionable labour initiatives such as the introduction of the infamous "Article 4" (summer 1984) or the legal changes concerning the National Federation of Labour (1986) are good examples of this phenomenon.

73. *To Vema.* 29 July 1984 and *Eleftherotypia* 30 July 1984.

74. D. Sellou, "To PASOK Meta to Synedrio" ("PASOK After the

Congress") *Fylladio* No. 10, September 1984.

75. *Exormise*. 7 June 1987 and *Socialistike Theoria kai Praxe*. July 1987, pp. 24-74.

76. The phenomena of resignations of M.P.s or top functionaries of New Democracy have been more than common. Many prominent figures of the party including former Prime Minister Rallis and cabinet ministers have resigned from the party. In fact some of them formed the Democratic Renewal (Democratike Ananeose – DEANA) in 1985 whose popularity however is far from threatening to New Democracy.

77. The ten per cent decline in the Party's electoral appeal in the last election (9.89%), which is its first electoral decline since its legalization in 1974, is but a further case in point of the party's sorry situation.

78. Only in the 1985 election has the party managed to win one seat in the national parliament.

7

Conclusion

The foregoing analysis leaves little doubt that PASOK's creation was the result of three major factors: the socio-economic development of Greece in the post World War II era and the politics to which this development gave rise in the mid-1960s; the resistance organizations that arose during the dictatorship; and finally. the idiosyncrasies and of the charisma of Andreas Papandreou.

In the immediate post war period, the working people of Greece were in effect kept at the margin of the political arena. Political mobilization was conducted through the old clientelistic structures, while the state, using the ghost of the civil war and the security valve of emigration, managed to keep social unrest under control. However as the economic and demographic map of the country changed, so did the population's attachment to the system of clientelism and the manner of conducting politics in the country as a whole. Rapid urbanization and the growth of the urban working class meant that these old forms of political articulation were becoming inadequate and outdated, particularly in the cities, while in the countryside they were adapted and modified. Overall however, there was a strong tendency for mobilization against the old restrictive practices and the repressive political apparatus. These factors led to the rise of George Papandreou's Center Union Party, the support for which was based on the growing belief that the problems of the country could be overcome through better planning and a strengthening of the country's democratic institutions through constitutional guarantees.

It is in this period that we must locate the conception of PASOK. The imposition of the dictatorship came about as a re-

Conclusion

sult of the instability in the power bloc which the growing radicalism of the sixties had created. And in fact, the resistance groups that sprang up in response to the dictatorship did not change the nature of the sixties radicalism. On the contrary they strengthened and consolidated it. If there was a common denominator among the numerous resistance groups and organizations, it was the rejection of the old configuration of politics and the need for the creation of a new political expression(s). PASOK was the creation of a number of left and left of centre resistance organizations, and especially Papandreou's own PAK. There is no doubt that PASOK will always carry the birth-marks of the anti-Junta resistance movement.

The undisputed charismatic personality of "Andreas", as the people of Greece soon learned to call him, for historical reasons and for reasons attributable to his own brilliance, became the culmination of all the pre- and post-dictatorship political currents. To begin with, Papandreou was able to best express the anti-clientelist, anti-old politics sentiments of the population, for he was part of the old political guard, but definitely not main-stream, as he had come into conflict with the hard core of the *palaiokommatikoi* during the days of the pre-Junta Centre Union. Thus at least symbolically, Andreas, more than anyone else, represented both the break away from and the continuity with the old politics that the tendency most conservative with regard to clientelism (especially in the countryside) seemed to favour.

Furthermore, Papandreou appeared to have all the qualifications necessary to respond to the technocratic political tendency which had gained some ground during the dictatorship as a number of anti-Junta activists had spent time in the graduate schools of Europe. Papandreou was an economist, a university professor of international acclaim, which almost by definition brought him closer to the middle class intellectual activists, who more than anyone else, had a technocratic orientation. Finally, Papandreou's association with the resurgence of Marxism in radical economics in the late 1960s and early 1970s appealed to the independent left radicals who had appeared in the *Anendotos* years and had gained some strength and sophistication during the resistance years.

These characteristics of Papandreou not only placed him as the undisputed leader of PASOK but also made the Movement a pole of attraction for a wide range of political currents which

Conclusion

were in fact one of the important factors influencing the development of the young Movement. Other factors of some significance include: the petty-bourgeois nature of the Greek social formation; the overall impoverished discourse of the Greek political culture; and its political inheritance from the resistance groups. The articulation of the various political currents was nothing other than the political expression of the social conflict by the newly born Movement. PASOK's organizational microcosm became the arena for the resolution of the conflict between the political polarities of the Greek social formation.

Thus it was not by accident that the Movement's structure became the focus of conflict between these political currents. But, given that none of these political currents could claim a numerical superiority, the battle over the organizational question, and not over the question of the Movement's policies became the key and the motor of PASOK's development. That is not to say however, that policies and political orientation are not intertwined with the organizational question but in the case of PASOK the actual internal debate and conflict over policies took place implicitly while the organizational ones were apparently dominating. The peculiarities of the various factions within the Movement gave PASOK's leadership (namely A. Papandreou and a few committed faithfuls who interestingly enough had come out of the three tendencies) considerable flexibility and automony *vis-a-vis* the Movement's internal conflicts. Thus it appeared that it was the leadership's deliberate choice to give to the Movement an orientation which was far removed from the radical premises of 1974, without at the same time endangering its unity and growth. Of course these were not the only reasons for PASOK's development in such a way. The overpowering strength of Papandreou did not grow within a political vacuum. PASOK's eventual "conservative" development was favourably conditioned by social and political factors external to the Movement.

To begin with, the Greek social formation, by virtue of its origins, is overpowered by what we will call the petty-bourgeoisification of almost all popular strata. This by no means implies that the Greek popular strata are objectively petty-bourgeois. Rather it means that the traditionally strong old and new petty-bourgeoisie, the close ties of the new urban working class with the peasantry and the institution of small ownership, the large peasantry (though in decline) and finally

Conclusion

the lack of a long industrial working class tradition gives an overwhelming petty-bourgeois flavour to Greek politics. This trait was/is primarily expressed through the short-sighted individualistic, acquisitive and conformist attitudes of Greek society. That meant an overall anxiety to capture power even if it had to be done in a sloppy manner which would overlook "meaningless details" such as party democracy and mass participation in policy formulation.

In addition, PASOK's development was the result of the political problematic of both C.P.s, the overall impoverished political discourse of the Greek political culture, the lack of a strong left-wing intelligentsia, and the disarray in the ranks of the dominant right-wing party (New Democracy) especially during the five years immediately following the departure of K. Karamanlis from its leadership. All these factors came to the assistance of those tendencies which eventually took the upper hand in the organization: the *palaiokommatikoi* and the technocrats.

Not only did the C.P.s not manage to develop any coherent and convincing criticism of PASOK's organizational and political developments, but they also failed to develop an attractive and threatening alternative to PASOK's increasing popularity. The political discourse and culture was also conducive to PASOK's eventual development. Despite appearances, Greek people relate to politics in an obsessive and highly unsophisticated way, much as we relate to our football teams. The Greek demon for politics rarely goes so far as to debate or contrast political claims and policies. Moreover, references to politics are rarely if ever connected to social issues or social classes. In addition, political parties have made virtually no serious effort to transform the Greek mass from voting objects into active, participating subjects in the political process. Thus the development of the rather centralized structure by PASOK grew almost naturally within this atmosphere. In addition, the fact that Greece, for reasons rooted in the structure of the educational system and in the nature of the left tradition in the country (i.e. many years of underground activity, strong Stalinist tradition), lacks a strong radical intellectual tradition was also conducive to PASOK's development. The absence of a radical intellectual analysis capable at least at the theoretical level of challenging PASOK's leadership's shakey theoretical claims, assisted in the "conservative" modification of the Movement's founding premises.

Furthermore, PASOK's roots in the resistance from 1967 to 1974 continued to influence its development. None of these resistance groups could claim either mass support or overwhelming popularity. Consequently, the party created out of these organizations was the result of the work of a small number of activists and not of mass political processes. And where the people are speechless, the demagogues are free to act. By definition, the absence of mass participation in PASOK's creation gave an edge to the strongest personality within this small group of people who took the initiative to create it. That person was of course none other than A. Papandreou. Thus PASOK naturally inherited many of the characteristics of the resistance organizations which established it. These were a wide spread voluntarism, almost a natural phenomenon among small groups of the kind, and a tendency to define politics and political strategy in a vague, scattered, and unfinished manner. In fact, it was the combination of these two characteristics of the resistance, as they were bequeathed upon the newly formed Movement, which contributed significantly to the admirable unity of the diverse groups that created the Movement in the first place. The vague and unfinished nature of the "3rd of September" was almost standard political practice for PASOK's founding members. After all, their voluntarist understanding of political action was a further guarantee of the of the final vague definition of the New Movement on their own terms.

But if the process of PASOK's creation and the characteristics of the political discourse of the resistance partially explain the unity among the establishing tendencies of the Movement, it was the 1974 general election which took place only two-and-a-half months after the "Declaration of the 3rd of September" which clearly initiated a differentiation in the balance of power between these groups. Thus as early as 1974, PASOK was almost forced to take a move towards electoral efficiency. This was something which under the circumstances could only lead to a concentration upon "pragmatism" and "efficiency", both of which were very far removed from the Movement's original premise of developing along the lines of a mass, participatory and democratically defined organization. The election became the first political project of the new Movement and it was inevitable that it would give an advantage to the organization's more astute political forces which in the context of the electoral com-

Conclusion

petition were the *palaiokommatikoi* and Papandreou.

The series of splits that followed the 1974 election highlighted the crises which resulted from the internal reactions of the various sections of the membership that objected to the Movement's new direction, and who were potentially a threat to its development into an in effect electoral, and centralized fashion. The practices which were established during PASOK's internal crises consolidated a new *modus operandi* within the organization. There was not of course a complete destruction of the political tendencies of the organization not entirely conducive to the Movement's electoral orientation, but rather a containment, so that they could not be obstructive to the pace of the organization's march to power.

In fact the existence of just these tendencies within the Movement was crucial to the success of the strategy which the leadership had chosen for the Movement. PASOK's strength was based on the peculiar political pluralism which originated among those groups that had created the Movement. The destruction of this pluralism would have meant an undermining of the organization's growing popularity. Thus the leadership, and especially Papandreou, who appeared to be the initiator of the Movement's new orientation, "cleansed" the organization's ranks of the most obvious opposition in order to speed the transformation of the Movement. But at the same time, and more importantly, the leadership tried to maintain the pluralist image and the facade of PASOK's original promise. Thus in the absence of any real patterns of internal communication structures, and thanks to the overwhelming power of the leadership, the organization managed to develop one language and practice within the organization through which the membership was able to believe in just about anything, and another for its public appearances and image. The "synthesis" of this two sided (two speed) practice of the Movement allowed the maintenance of the necessary facade for the Movement's public (and pluralistic) image, since the public practice of the Movement was by definition more important. This of course constantly gave an advantage to some tendencies of the organization, namely the *palaiokommatikoi* and at the same time undermined its "radical" current. And interestingly enough, the more the premise of participation was translated into electoral hyperactivity, the more buoyed up with hope the membership became, supporting the idea of a strong organization, and failing to see at the same time

Conclusion

that the true winner was parliamentarism.

In this way, PASOK's leaderhip appeared to keep the balance within its organization while heavily favouring the electoral direction of the movement. When by 1977 it had become clear that it was only a matter of time before PASOK would move into government, the leadership started to openly favour the technocratic tendency, which up to that point was visible but definitely wielding no particular weight. The "fear of power" was vital in making the technocrats a powerful part of PASOK's internal pluralism. But, the technocrats could not claim omnipotence, since given their political inexperience they needed the *palaiokommatikoi*, while in turn the latter needed them to prove to the electorate that they could get the job done. Finally, both needed the party structure (in which the left tended to predominate) in order to win the election, as well as to mobilize support for the Government's policies.

This was the essence of the balance of power within the Movement's structure when PASOK triumphantly entered the government. As was expected, the Movement transferred its internal affairs, manners and politics into the governmental apparatus. Its trilateral political tendencies were soon to turn the governmental apparatus into the arena for their reproduction. If anyone is to understand the erratic, and often apparently contradictory nature of the Socialist policies, it will be necessary to understand the checks and balances between PASOK's internal tendencies. Once in government, the technocrats were essentially given an objective advantage when they were awarded the most influential economic portfolios. However, the "jerky" fashion through which the Government seemed to pursue its economic strategy as well as the Keynesian intervals indicate the Government's response to the pressure exerted by the politically influential *palaiokommatikoi*. The cost-free social reforms can in turn be seen as the Government's response to the admittedly much less influential radical tendency.

Every government though, has to "deliver the goods" within the parameters of its given domestic and international economic resources. PASOK's amazingly rapid and almost circumstantial way of reaching power through parliament, guaranteed that the actual implementation of policies was to be left to the experts. It did not take long for the market to force the "Keynesian hopes" into retreat and to give rise to the monetarist policies of austerity which in effect brought the technocrats to the fore. It

Conclusion

has in fact been the bitter reality of governing a country with such a schizophrenic economic and social existence that has gradually led to the establishment of the technocrats as the dominant faction within Papandreou's government and party. From the summer of 1983, when the controversial anti-labour legislation "on socializations" was introduced until the fall of 1985 when the severe economic measures of austerity were introduced that the Government not only demonstrated a determination to go ahead with its economic strategy but also indicated that the technocrats had almost completely taken over. This in fact severely undermines the traditional pluralistic political pillars of PASOK and in turn imposes critical questions about the Movement's future.

PASOK's pluralism, along with its charismatic leader, was as we have seen, the basis of its popularity and strength. If the domination of the Greek government party by the technocrats continues (and all things being equal, it will), then the Greek Socialists will not only become more consistent, but also more autocratic, given the nature of their technocratic political orientation. At the same time however, this enforced Bad Godesberg and modernist homogenization will almost mathematically lead to PASOK's own demise. After all, the undermining of the *palaiokommatikoi* does not only mean that a number of old politicians might lose their seats in parliament, but also and primarily, that the interests of the strata which base their social reproduction on small ownership and the numerous parasitic small enterprises will be undercut. In turn, this means a severe undermining of PASOK's political support, which to a great extent is based on the strata which the technocrats are trying to undermine. Likewise, the benching of the radicals of the party, in spite of the fact that they have very limited direct links to the social base, minimizes (if not eliminates) the capacity of the Government to maintain the somewhat radical image necessary to continue to draw support from the radicalized new urban working class.

PASOK's odyssey has ended. After over ten years of often puzzling developments, the Movement has been forced to land in Ithaca. It had to capture power and confront the harsh reality of actually running the country before making its political orientation clear. Under these considerations, it seems to us that the "peculiar" and "unique" phemonenon of PASOK is reaching its limits as it has been forced to undermine the very basis of its

Conclusion

strength: the tacit articulation of a tri-polar pluralism. Of course there is another factor which is going to determine the timing and probably the actual outcome this scenario. This is none other than Papandreou himself. Papandreou has, up to this point, masterfully managed to capitalize upon and articulate in a unique fashion, the politics of all the currents of his party. In fact, as we have seen, the reproductive strength of these tendencies depend on him much more than on anyone else, since up to that point, it seemed that no one of them could become dominant by itself. Even today, when everything hinges on the technocratic omnipotency, it is Papandreou's open support for their policies, which in turn has given the latter their strength. After all, without Papandreou there is no PASOK. Therefore Papandreou's own consistency in support of the technocrats and most of all, the extent of his capacity for using his charisma to mobilize support for his Government's policies will determine the timing of PASOK's forced suicide.

Throughout this study, we intentionally avoided following any of the conventional terminology and labeling PASOK a "social-democratic," "socialist," "bourgeois," or any other type of party. These are categories which, if not historically and concretely defined, run the risk of becoming meaningless and misleading generalizations. Categories such as these can thus be used at best as approximations or abbreviations. But in our case, after a study of several hundred pages, such simplistic approximations or abbreviations are hardly necessary. PASOK is the specific, and in this sense the unique, expression of the all the factors of the Greek social formation which were discussed above. To label it is to undermine its specificity.

Papandreou's key role in Greek politics of the 1980s has some theoretical implications concerning social change and historical developments overall. While his significance is undoubtedly the product of conjunctural considerations both in and outside his party, there are equally good reasons to believe that his actual style and often Bonapartist practices *vis-a-vis* his party or the government are the results of long standing practices customary to almost all significant political figures in the Greek party system. It appears then that societies have a tendency for inertia in their practices. This Bonapartism, for example, primarily as introduced by General Papagos after the civil war and thereafter masterfully practiced by Karamanlis was a byproduct of the "guided democracy" of the 1950s. There is no

Conclusion

doubt that in the 1980s the material base of the "guided democracy" is long gone. However, the practices surrounding the role of national leadership, though very much altered and transformed have basically survived.

We can draw similar conclusions from the constant metamorphosis of the institution of clientelism that Greek politics are so famous for. Although the original material conditions of the institution are long gone and PASOK's organizational structure, despite its shortcomings, has greatly undermined this traditional institution, the "green guards" are the living example of its persistence. Historical phenomena have a remarkable endurance and a surprising autonomy from their original material base. This is something that "materialist" optimism or radical voluntarism tends to undermine. If our forecast about the future demise of PASOK is correct, then future endeavours towards left political mobilization in Greece should seriously consider the heritage of Papandreou's party.

There is a further point which we hope this study has made clear, i.e. that a full and comprehensive study of governmental or party policies necessitates the analysis of party structure. Contrary to the research habit in the field, in which parties are usually examined solely ideologically, politically, or in terms of their social base, the question of organization and party structure should be an intrinsic part of any research on parties. The same thing goes for any inquiry concerning the policies of a party in power. However it seems that studies of government policies rarely go beyond the state, political economy, ideological or leadership considerations.

In order for policies to be developed and articulated, for the leadership to be recruited, for even social/class representation to be achieved, an organizational structure is necessary. Party structure is the material reason and base of all other aspects of party existence. As such, it conditions the political/ideological dimension and the class participation in political parties just as it is conditioned by them. Consequently, party structure is vital to an understanding of a party's policies and moreover, to the dynamics of that party. This is more so when a party moves into power, since the party carries with it its structure and organizational customs and patterns. A failure to recognize the importance of the party structure in this way will result in the jeopardizing of the possibility of understanding a government's policies and outlining the dynamics and the actual future of that

Conclusion

party. PASOK makes an ideal case-study in that regard.

Finally, one of the useful, but hardly from the first conception anticipated, aspects of this study is that it constitutes an exceptional case-study of the role of the individual in history. Individuals with special qualities have the capacity to become real factors in historical developments. However, these "special qualities" are not metaphysical capacities that an individual possesses by birth or otherwise. Rather they are the historical or circumstantial capacity of the particular individual to crystallize in his/her personality the historical interests of certain social classes and strata. Therefore, the charisma of the exceptional individual is not inexhaustible, since it depends on that person's capacity to capitalize, express and serve these interests. As soon as this special relation, for whatever reason, vanishes, the charismatic personality loses its extraordinary capacity to "make history". This it seems will also be the case with Andreas Papandreou. His future as a leader of PASOK will depend of his capacity to continue his popular appeal to the Movement's social base and his organization's political tendencies, as well as on the failure or the success of these interests and concerns to find other forms of articulation. This of course will not only determine Papandreou's future, but that of PASOK as well. It will not be long before Papandreou's capacity is critically tested in this way. The militant mobilizations in response to the Government's austerity measures, and the expanding influence of the right (municipal elections 1986) are cases in point. The answer to this decay of PASOK cannot come from within its organization. If anything, this study has shown that the Movement, in its short march to power, has destroyed the organizational structures which might have allowed the possibility of a radical revitalization of PASOK.

To conclude, despite its departure from its original radical promise, its conservatism and its overall shortcomings, PASOK is not to be dismissed lightly, even from a democratic socialist point of view. The Movement's radical promises, although far from being materialized, have made their imprint (admittedly in a vague and confused fashion) upon the collective memory of the Greek people. The shock of austerity and statist authoritarianism, seem to have ensured that the latter is not about to be struck with amnesia. The PASOK experience has undoubtedly been a harmful one for the credibility of a democratic socialist promise, since it increasingly associates radical/socialist

Conclusion

and participatory rhetoric with policies of austerity and profoundly undemocratic party and state policies. There is however another, more positive side to the story created by PASOK's own initial radical promise, as the discrepancy between this promise and its actual practices open new possibilities for democratic socialist action.

It appears that democratic socialism in Greece, as everywhere else for that matter, has a rather tenuous existence. However, no one claimed that the radical transformation of capitalism would be either a play of simply one act, or a tea party.

Appendix I

Declaration of the Fundamental Principles and Aims of the Panhellenic Socialist Movement*

The tragedy of Cyprus as well as the dangers which have resulted from both the unhesitant, expansionist policy of the Pentagon within the framework of NATO and the attempts of the American backed Junta to transform our armed forces into an exclusive policing instrument of the Greek region dominates the mind of every Greek. However the unity of the people around its decision to uncompromisingly confront the external danger and every threat against the integrity of our national presence, does not justify the inaction of the government in three areas: the punishment of the parties guilty of the seven year long dictatorship, of the massacre of the Polytechnic (School), and of the Cypriot tragedy; the clearance of the state apparatus; and the all around rehabilitation of the occupation victims. There is deep concern among Greek people because the Government's declarations for the re-establishment of normal political life will be only promises if they are not accompanied as soon as possible by punishment, purification and rehabilitation. It is said that the time has not yet come. This is the prevailing national issue. But the argument does not hold. It is said that the time has not come. How is it possible for those who are responsible for the national disaster to remain in positions crucial to the nation? Precisely because Greece is at a critical stage, precisely for this reason, we have to pursue with courage the punishment, the purification and the rehabilitation, (in order) to protect the nation, to open the road which leads to unbounded popular sovereignty and democracy.

Today, within this framework, we must interpret our decision to pursue a political action, a declaration of the fundamental principles and aims of a new political Movement: the Panhellenic Socialist Movement. Only through the active political presence (and) participation of citizens throughout Greece, will both our national independence and popular sovereignty be guaranteed. The time has come to pass from passive expecta-

Appendix

tion to the future of our country.

The root of the misfortune is the dependency of our Country. The seven dark years which passed with the gloom military dictatorship and the tragedy in Cyprus are nothing but a particularly crude expression of Greece's dependence upon the imperialist establishment of the USA and NATO. Greece has been transformed into a front line outpost of the Pentagon to more efficiently save the military and economic interests of big monopolies. The state apparatus, the armed forces, the (political) parties, the trade unions, the country's Political leadership had been eroded to such an extent that the establishment of a foreign backed military dictatorship was made possible, when it was decided that this would be in Washington's interests. The military dictatorship was established in order to stop our people's march towads popular sovereignty and national independence. The coup against Makarios took place and the brutal Turkish invasion of Cyprus followed, so that the island could be bisected and finally became a new military outpost for the USA and NATO in the eastern Mediterranean.

Our fatherland has been transformed into an unfenced vineyard, so that our economy has been eroded by the multinational companies of the USA and the West (of course) always in cooperation with indigenous commercial capital. So the Greek countryside wilted, and the sweat of the peasants became unproductive, emmigration (became necessary) as well, surplus labour in the capital and abroad in Europe, Australia, and Canada, continued.

The march towards subjugation, the mining of our national interests, the corosion of popular sovereignty, the economic decline and the exploitation of the Greek wage earners must be stopped. On the contrary, we should pursue with courage and decisivenes, the founding of a new Greece.

We are announcing today the beginning of a new political Movement which we believe expresses the desires and the needs of the ordinary Greek, a Movement that will belong to the peasant, to the worker, to the artisan, to the wage earner, to the employee, to our courageous and enlighted youth. We invite them to fill its ranks; to stalk and participate in the direction of a Movement which will simultaneously push forward our national independence, popular sovereignty, social liberation and democracy, in all aspects of public life.

The basic, fundamental goal of the Movement is the creation

Appendix

of a polity free of foreign control or intervention, a polity free of the control or the influence of the economic oligarchy, and devoted to the protection and the service of the people. National independence is inseparably connected with popular sovereignty, with democracy in every aspect of the country's life, with the active participation of the citizen in every decision that concerns him. Furthermore, it (national independence) is interwoven with the disengagement of our economy from the control of foreign monopolies and local, comprador capital which shape our economic, social, political and cultural direction not according to the interests of the people but according to those of the economic oligarchy. Of course Greece has to withdraw both militarily and politically from NATO, (and from) all the bilateral agreements, which have allowed the Pentagon to transform Greece into an outpost for the advancement of its expansionist policies. But behind NATO, behind the American bases, there are the multinational monopolies and their domestic subsidiaries. For this reason, social liberation and socialist transformation have become the founding stone of our Movement. (This will ensure...) ...that the farmer will enjoy the product of his sweat and his land, so that the worker, the artisan, the wage earner, the ordinary Greek will enjoy the product of his pain. ...that the striking incomes inequalities between geographical regions and among social strata, which characterize modern Greece, can effectively be fought. ...that the exploitation of man by man will be eliminated. ...that the people will actively participate in the planning of the economic, the social and the cultural course of the Country. ...that work and housing will be guaranteed for all Greeks. ...that the privileges of the few in medical, hospital and pharmaceutical care will be abolished. ...that mother, child and the aged will be protected. ...that the social and the economic equality among the two sexes will be fortified. ...that thought will be freed and (that) education will become an estate of all Greeks.

Today's declaration of the principles of the Panhellenic Socialist Movement is for the establishment, the recruitment and the foundation of a Movement that we want to become the agent of all genuinely progressive and democratic forces of the country. We ask these forces to unite to pursue the struggle. The fundamental principle of the Movement is the absolutely fortified democratic procedure — from the rank and file up to the top leadership — with absolute equality for all the members who will be recruited into it. Also the programme and the organizational form will be co-decided in the course (of events) with equal participation of all the members of the first congress which will be called soon. And this (will be pursued) within the framework of a guaranteed democratic procedure.

Our people have had a bitter experience with the party forms of the past which were based on a feudal relationship between leaders and deputies, between deputies and "kommatarches", between "kommatarches" and voters. (The people have had a bitter experience) with party mechanisms which have replaced the principles, the programme and the democratic procedure with favouritism and backstage activity. The demand for principled political organizations, which will be distinguished by the free (and) democratic expression from the base, so that the leadership will be bound to the political decisions, and so that there will be consistency and continuity, is universal.

We are certain that our declaration today is a declaration which mirrors the beliefs, the desires, the demands, the vision of the Greek people. It aims to generate discussion and discourse at the national level. Our declaration today becomes the compass which will guide our march towards a new, reborn, human, and democratic Greece, a Greece which will belong to the Greeks.

The Panhellenic Socialist Movement is a political Movement which fights for the following aims:

>National Independence
>Popular Sovereignty
>Social Liberation
>Democratic Procedure

The struggle of the Panhellelenic Socialist Movement for our national rebirth, for a socialist and democratic Greece, is based upon the principles that our national independence is the prerequisite for the realization of popular sovereignty; that popular sovereignty is the prerequisite for social liberation; and that social liberation is the prerequisite for the realization of political democracy.

For the Panhellenic Socialist Movement, the military regime of junta, which was imposed by the coup of 21 April 1967, was nothing but special brutal expression of the colonization of Greece by the Pentagon and NATO with the collaboration of the dependent western European and domestic commercial capital. Its aim was the accommodation of the strategic and economic aims of American capital in the eastern Mediterranean. For this reason the struggle of our people aims primarily at the definite elimination of the causes which led to the brutal, seven year long dictatorship. This struggle cannot be justified unless there is:

a) Punishment of the guilty (parties) and of the torturers of the seven year long junta (regime) and of those responsible for the treason in Cyprus.

b) A complete rehabilitation of the victims of the dictatorship.

c) An immediate invalidation of all the servile and oppressive special measures of the junta as well as of similar bills of the pre-dictatorship governments.

d) The securing of the free return of political refugees to the fatherland.

e) The purification of the entire state apparatus.

f) The abolition of the para-state and of the partisan one.

g) The immediate placement of the armed and security forces in the service of the Nation and the People and their subjugation in effective, complete and continuous contol by the legally elected political leadership.

For the elimination of the system which led to the imperialist occupation of our country and of the conditions which created it, maintained it and protected it, so that the foundation of a genuine, non-monarchical, reborn and socialist Greek Democracy (will be established), the Panhellenic Socialist Movement puts as a prerequisite the realization of the following concrete goals:

1. Power springs from the people, expresses the people and serves the people. The social, economic and political power structure in our country is articulated in a way which excludes the violation of the popular will in any fashion.

2. The right of defence of every citizen against any attempt of the abolition of legal power, the abolition of the constitution and the enslavement of our people is guaranteed.

3. For the basic rights of the citizen, the civil rights charter of the UN is valid. Freedom of opinion and expression, freedom of association for the achievement of collective goals within the framework of the constitution, the inviolablity of individual rights, are not only guaranteed but are also protected effectively

by Justice which is independent. The Greek citizenship is not to be removed.

4. The constitutionally guaranteed social and economic equality of the two sexes is guaranteed.

5. The direct and active participation of all citizens in the political life of the country is secured with genuine democratic procedures.

6. The right to work for all citizens, men and women is constitutionally secured.

7. Unionism is freed from dependency upon the economic oligarchy and the curatorship of the state, (it) is safeguarded as a free and autonomous movement and is put in the service of the working people.

8. The church is separated completely from the state and the monasteries' property is socialized.

9. Greece disengages itself from the military, political and economic alliances which undermine our national independence and the sovereign right of the Greek people to plan for itself the social, economic, political and cultural march of the country.

10. Greece follows a dynamic independent foreign policy with the (following) goals: the guarantee of the country's integrity, the safeguarding of the unbounded people's sovereignty and the realization of the aims of the Greek people. As a country which exists simultaneously in Europe, in the Balkans and the Mediterranean, it makes its presence felt in these three regions. The nuclear disarmament of the Mediterranean and Balkan regions, the neutralization of the Mediterranean from the military blocs, the tightening of the economic and cultural relations with the people of Europe and of the Mediterranean, as an offer to international peace, the fraternity of the people and the structure of all countries towards an all human (pananthropist) and universally free Community with equal treatment and equal rights among all people, are her (Greece's) permanent goals.

11. The international contracts and agreements, which have led Greece to an economic, political and military dependency upon the monopolistic conglomerates of the West and particularly of American imperialism become invalid.

12. The social liberation of the Greek working people, which in the long run is identified with the socialist transformation of the society is to be pursued. The prerequisite course towards for today's visible future is:

Appendix

a. The socialization of the financial system in its entirety, of the basic units of production as well as those of big domestic and foreign traders. Simultaneously, the inclusion of the agricultural enterprises into new forms of co-ops, will be promoted systematically with activities which will be expanded to the supply of raw materials and to the processing, packaging and distribution of their products. These organizations will eliminate the middle-man who exploits the product of the sweat and of the land of the farmer. Also in the handicraft (sector) co-ops will be promoted.

b. The regionally decentralized social planning of the economy, which is combined with the control of the productive units by the employees (i.e. with self-management) and by the appropriate social couriers. Appropriate couriers are the state, the region, the city or the community, to depend on the size, the type and the significance of the productive unit.

c. Administrative decentralization with the strengthening of municipal power.

d. The systematic and progressive narrowing of the gap between the higher and the lower incomes by region and profession.

e. A housing and city planning policy which will secure civilized housing for every Greek family.

f. A new education (system) so that the barriers which prohibit the expansion of knowledge will be abolished and so that freely thinking and socially responsible citizens will be created. Education is the responsibility of the social collective. Private education is abolished. Free and compulsory education is secured for all Greeks without exception; the educational policy, which is introduced secures the wide participation of all popular strata, as well as the participation of students in the planning of education and the administration of the educational institutions.

g. The socialization of health which means free medical, pharmaceutical and hospital care, preventative health care for all Greeks, the abolition of private clinics and all privileges in providing medical and hospital services.

h. A system of social security for health, accidents, old age and unemployment which is extended to all Greeks.

i. The protection of mother and child.

j. The protection of the environment, the betterment of the quality of life in combination with making our national and popular traditions worthy, and the participation of all people in cultural development.

The economic, political, social and cultural goals of the Greek working people — workers, farmers, wage earners, employees, youth, small-professionals and artisans — (which are) the foundation of a society without alienation and bureaucracy, will be realized with a continuous popular watchfulness, contol and mobilization.

The Panhellenic Socialist Movement calls upon the Greek People to get organized in rank and file organizations, to participate directly in the further formation of its programme, in all decision making and in the appointment of its functionaries at all levels. So we will continue the stuggle for an independent, socialist and democratic Greece with a new intensiveness and decisiveness.

The Framework of the Organizational Structure of the Panhellenic Socialist Movement

The Panhellenic Socialist Movement is organized on the basis of genuine democratic procedure at all levels of its structure.

In the first phase, a Provisional Central Committee will be formed which will promote the organization of the Panhellenic Socialist Movement from the rank and file and up and it will function as its higher administrative body until the organizing of the first Panhellenic Congress.

The Provisional Central Committee will be organized immediately in Subcommittees of Action and will proceed directly with the appointment of provisional, regional, provincial and local committees which in their turn will go on with the organizing of the people who agree to serve the principles (of PASOK) within rank and file organizations. These organizations with their active participation will contribute to the further development of the programme of the Panhellenic Socialist Movement as well as in the appointment of its functionaries at all levels, within the framework of genuine democratic procedures.

After (the rank and file clubs) have been democratically structured, provincial and regional congresses, and finally the Panhellenic Congress of the Panhellenic Socialist Movement, which will be the highest body of the Movement will be organized. Only democratically elected bodies will participate in

Appendix

the congresses. The Panhellenic Congress of the Panhellenic Socialist Movemnt is the highest body of the Movement and its decisions bind all levels of its hierarchy. The Panhellenic Congress elects the Central Committee of the Panhellenic Socialist Movement.

The Provisional Central Committee will produce (a) a constitution, (b) a constitution of the internal functioning of the (various) bodies, and (c) a programme of action which it will submit to the Panhellenic Congress for approval.

The Panhellenic Socialist Movement with the wide participation of the people, aims at becoming the vanguard in the formation of a new political Movement. This Movement, away from the traditional political schemas and historically dated procedures, will become the organizational carrier which will give expression to the general will and the political, social and cultural pursuits of our people.

3 September 1974

---

* In the translation we have tried to follow as closely as possible the expressions, the structure and the rhetoric used in the original. We believe that the form of the language used is not divorced either from the content or its user or (perhaps most importantly) from its welcoming audience. The reader will often be puzzled by the awkwardness, the convoluted nature as well as the overall poverty of the language used. We believe, that the problem lies with the original document.

Appendix II

The Central Committees Of PASOK

PASOK's First Central Committee – 10 October 1974

This 75 member central committee was *appointed* and announced by Andreas Papandreou on 10 October 1974.

Alexopoulos, Thanases
Androutsopoulos, K.
Bardakis, K.*
Bases, Laokr.*
Basileiades, Damianos*
Basileiou, G.*
Boyiatzides, Demetre
Boulgare, El.
Giotas, S.
Dalavagas, G.
Dedes, Sot.*
Demopoulos, N.
Drosoyiannes, An.
Eleftheriades, Demetres
Efthumiou, P.*
Ziagas, Michalis
Zoubogiorgos, N.
Zigoyiannes, G.*
Thomidou, S.*
Kabounides, Sp.*
Kavoliotou, Christos
Karagiorgas, S.*
Karras, Ant.
Katres, G.*
Katsifaras, G
Kessides, Pan.
Kissonas, G.*
Kokkola, Ag.*
Lampropoulos, O*
Livanes, A.
Merkoures, Sp.*
Micha, Maria
Manolkides, K.*
Michalopoulos, Nikos
Bournazakis, N
Dolkas, L.*
Xanthakis, N*
Xydes, Al.*
Ekonomopoulou, M
Pakos, F.*
Pantazopoulos, St.*
Papanikos, D.
Papadopoulos, Ap.*
Papathanasiou, Christos *
Papandreou, Vaso
Papadatos, Andreas
Polites, Andreas*
Prokos, Ap.
Pyrzas, K.*
Rokkos, D.
Sipetanos, G.*
Sakellapopoulos, A.
Stangos, A.*
Stavrides, N.
Seitanides, N*
Semites, K.

Appendix

Kourte, Chr.
Koutsoyiannes
Konstantopoulos, N.
Kotsakis, N.*
Lagakos, Demetres*
Lazarides, Pan.*
Laliotes, K.
Lianes, G.*
Filippatos, L.*
Chousanthes, Yiannes

Skoula, Tasia
Tsouras, Thanases
Tsochatzopoulos, Akis
Tzortzes, Yiannes*
Filias, Bas.*
Fragias, Andreas
Fleming, Amalia*
Filippou, Katerina
Charalabopoulos, G.*

* These members resigned, withdrew, or were kicked out of the Movement within the first year of the Movement's life.

Central Committee Elected By the Pre-Congress, 16 March 1975

Fleming, Amalia*
Valyrakis, Sifes
Papandreou, Vaso
Livieratos, Demetrios
Pantazopoulos, Stelios
Laliotes, Kostas
Intzes, Tasos
Avyerinos, Paraskevas
Nikolaou, Kostas*
Stavrakakis, Menas
Michalopoulos, Nikos
Karras, Antones*
Konstantopoulos, Nikos*
Zoubogiorgos, Nikos*
Nikolaides, Pavlos
Kissonas, Giorgos*
Prokos, Apostolos
Veryvakis, Levteres*
Zafeiropoulos, Yiannes
Touloupas, Demetres*
Zigoyiannes, Giorgos*
Dragotes, Evthymios
Papadonikolakis, Yiannes
Lazarides, Panteles*

Semites, Kostas*
Tsochatzopoulos, Akis
Tsekouras, Yiannes
Merkouri, Melina*
Kedikoglou, Vasiles
Taxiarchou, Giorgos*
Papageorgopoulos, Takis
Dalavagas, Giorgos
Kirkos, Giorgos
Labrou, Petras
Micha, Maria
Kokkola, Angela*
Filias, Vasiles*
Tsouras, Thanases
Tzouvanos, Demetres
Notaras, Yerasimos*
Tsouyiopoulos, Giorgos*
Griniatsos, Giorgos
Nestor, Stelios*
Archontakis, Andreas*
Rokofilos, Christos*
Evthymiou, Petros*
Stagos, Asteres*
Markopoulos, Christos

Appendix

Manolkides, Kostas*
Kavounides, Spyros*
Tsigaridas, Kostas*
Papoulias, Karolos*
Koures, Kostas
Lampropoulos, Odysseas*
Papathanasiou, Christos*
Daskalakis, Giorgos
Thomidou, Soula*
Giotas, Stathes*
Barakos, Tasos
Vardakis, Kostas*
Athanasiou, Vasiles
Mastorakis, Giorgos

Vasses, Laokrates*
Staridas, Spyros
Fouras, Andreas
Papademetriou, Christos
Dolkas, Labes*
Rokkos, Demetres*
Fratzeskos, Fratzeskos
Mortzos, Yiannes
Tountas, Yiannes*
Kaloudes, Spyros
Tzoumakas, Stefanos
Livanes, Antones*
Gennematas, Giorgos*

* These members resigned, withdrew or were kicked out during the 1975 crisis. It is interesting to note that among these individuals we find Simites, a present senior minister of PASOK's government; Karras, a present senior political advisor to the government; and Livanes, the present director of the P.M.O. None of these ever officially submitted their resignation. Rather, they simply withdrew from the organization for a while. Gennematas, later senior minister of Government and member of the infamous troika, signed the collective resignation letter but remained silent on the other issues related to the actual events of the 1975 crisis.

Central Committee Elected by the Movement's Panhellenic Conference, 8-10 July 1977

Tsochatzopoulos, A.
Laliotes, K.
Gennematas, G.
Valyrakis, I.
Tzoumakas, St.
Nikolaides, P.
Rokkos, D.
Alevras, Yiannes
Semites, Kostas
Pavlides, Vangelles

Avyerinos, P.
Yiotas, St.
Christodoulides, A.
Papandreou, Vaso
Michalopoulos, Nik.
Daskalakis, G.
Markopoulos, Chr.
Charalabopoulos, Yiannes
Morales, Petros
Chras, Nikos

Appendix

Kostopoulos, Soteres
Morales, Petros
Exarchou, Giorgos
Lyberakis, Lyb.
Papathanasiou, Ant.
Papadopoulos, G.
Golfinopoulos, Thanases
Zacharapoulos, Vas.
Athanasoules, Thanases
Politopoulos, Kostas
Vales, Thanases
Petsas, Chr.
Kapatanopoulos, Th.
Sakellares, Mich.
Charalabides, Mich.
Ralles, Thanases
Nikainas, Tasos
Paidakakis, Yiannes
Karamanou, Anna
Mavrakis, Dem.
Pavlou, V.*
Kargopoulos, N.*

Gaitanides, Demetrie
Yiannopoulos, Vangelles
Mortzos, Yiannes
Ouzounides, Ares
Kouloures, Kimon
Karavele, Tzene
Dores, Michalis
Chatzes, Kivos
Souladakis, Yiannes
Laskarakis, Yiannes
Bakos, Kostas
Serafeimides, St.
Papas, Alekos
Soterles, Demetres
Papoutses, Christos
Kyriakou, Loukas
Ziagas, Yiannes
Zarras, Labros
Giokas, Yiannes
Fotopoulos, Th.*
Tziolas, L.*

* These members resigned from or were kicked out of the Movement for political reasons.

Central Commmittee Elected at the Movement's First Congress, 13 May 1984

Athanasoules, Athan.
Akritides, Nik.
Alevpas, Yiannes
Arsene, Maria
Avyerinos, Par.
Valyrakis, Sefes
Vounatsos, Dem.
Geitonas, Kon.
Georgakis, Char.
Yiannopoulos, Evang.
Givalos, Men.*

Akrita, Sylva
Alexandres, Stathes
Amalos, Tasos
Arsenes, Yerasimos
Vainas, Yiannes
Vgenopoulos, Nikos
Gaitanides, Dem.*
Yennematas, Georgos
Georgiades, Athan.
Giotas, St.
Golfinopoulos, Athan.

Appendix

Daskalakis, Georg.
Drosoyiannes, Ant.
Zakolikos, Pavs.
Ziagas, Yiannes
Thomas, Georg.
Intzes, Vas.
Kaloudes, Sp.
Karavele, Evyenia
Karamanou, Anna#
Kastanides, Char.
Katsibardes, Georg.
Kafires, Vas.
Kitsioules, Dem.*
Kosakas, Ant.
Kourmatzes, Athan.*
Koutsogiorgas, Menios
Kyriopoulos, Georg.
Kostopoulos, Sot.
Malesios, Evang.
Labrou, Petros
Mamalakis, Evang.
Manteles, Tasos
Matragas, Er.
Merkouri, Mel.
Michalopoulos, Nich.
Birdimires, Georg.
Moraetes, Georg.
Neonakis, Mich.*
Doutsos, Andr.*
Ekonomou, Pant.*
Paidakakis, Yian.
Palaiothodoros, Dem.
Papayiannes, Vas.
Papadionysiou, Vas.
Papaioannou, Milt.
Papandreou, Vaso
Papaspyrou, Yian.
Papoulias, Karolos
Piperyias, Dem.#
Priovolos, Vas.
Rapti, Sylvana
Reppas, Dem.

Daskalakis, Man.
Dores, Mich.
Zarras, Labros*
Thermos, Pan.*
Intzes, Anast.
Kaklamanes, Ap.
Kapetanakis, Thed.
Karakonstantakis, M.
Karras, Antones
Katsanevas, Theod.
Katsefaras, Georg.
Kissonas, Georg.
Kokkinovasiles, Chr.*#
Kouloures, Kimon
Koutseleou, Micha*
Kypriotaki, Maria
Konstantinou, Floros
Laliotes, Konst.
Labrake, Eirene
Laskarakis, Ioan.*
Manikas, Stef.
Markopoulos, Chr.
Mavrakis, Dem.*
Metropoulos, Al.#
Bazianas, St.
Bristoyiannes, Ger.
Morales, Petros
Nichalaou, Yian.*
Xarles, Georgios*
Pagalos, Theod.
Paipoutlides, Kon.
Panayiotakopoulos, Dem.*
Papadatos, Ioan.*
Papathanasopoulos, A.
Papanayiotou, K.
Papandreou, Georg.
Papastavrou, Makis
Papoutses, Christos
Pitakes, Ioan.
Rabavilas, Poan.*
Ravtopoulos, Georg.
Rizopoulos, Anast.*

Appendix

Rochos, Dem
Sakellares, Lyk.
Semites, Kon.
Skandalides, Kon.*
Smyrles, Chr.*
Spyropoulos, Rov.#
Strates, Antones
Soterles, Dem.
Trikoukes, Man.*
Tsimas, Kostas
Tsouras, Athan.
Floros, Yiannes
Chalatses, Theod.
Charalabopoulos, I.
Chatzemichalis, Fotis*
Chatzes, Fivos*
Christoforidou, Lila

Rysios, Alex.
Sapoudzes, Dym.
Siakares, Kon.
Skoularikis, Yian.
Souladakis, Yian.
Stavrakakis, Men.
Sfaggos, Kostas
Tzoumakas, Stef.
Tritses, Ant.
Tsovolas, Dem.
Tsochatsopoulos, A.
Fouras, Andr.
Charalabides, M.*
Charalabous, Yian.
Chatzenakis, Man.
Chatzopoulos, Yian.

* Those members who are the only ones who are not M.P.s or who do not hold governmental or other public posts.

These were trade unionists who were expelled from the Movement in the 1985/1986 crisis over the Government's economic policies of restraint.

Appendix III

Election Results

17 November 1974 Election

	% of Votes	Seats
PASOK	13.58	13*
New Democracy (Nea Democratia)	54.37	219*
Union Centre – New Forces (Enose Kentrou – Nees Dynames)	20.42	60*
United Left (Enomene Aristera)	9.47	8
Others	2.16	0
		300

* After the bi-elections of 20 April 1975. the distribution of seats in parliament became: PASOK 15, N.D. 216, E.K.-N.D. 61.

20 November 1977 Election

	% of Votes	Seats
PASOK	25.33	93
New Democracy (Nea Democratia)	41.85	172
Union of the Democratic Centre (Enose Democratikou Kentrou)	11.95	15
Communist Party of Greece (K.K.E.)	9.36	11
National Front (Ethnike Parataxe)	6.85	5
Alliance of Left-wing and Progressive Forces* (Symmachia Aristeron kai Prodeftikon Dynameon)	2.72	2
Neo-Liberal Party	1.08	2
Others	0.86	0
		300

* Included the Communist Party of the Interior (Kommounistiko Komma Esoterikou), E.D.A. (Unified Democratic Left), Socialist March (Sosialistike Poria), Socialist Initiative (Sosialistike Protovoulia), Christian Democracy (Christianike Democratia).

Appendix

18 October 1981 Election

	% of Votes	Seats
PASOK	48.07	172
New Democracy (Nea Democratia)	35.87	115
Communist Party of Greece (Kommounistiko Komma tes Elladas)	10.93	13
Communist Party of the Interior (Kommounistiko Komma Esoterikou)	1.34	
Progressive Party (Komma Proodevtikon)	1.68	
KO.DE.SO. (Party of Democratic Socialism)	0.70	
Others	1.41	

18 October 1981 European Election

	% of Votes	Seats
PASOK	40.29	10
New Democracy (Nea Democratia)	31.53	8
Communist Party of Greece (Kommounistiko Komma tes Elladas)	12.68	3
Communist Party of the Interior (Kommounistiko Komma Esoterikou)	5.15	1
Progressive Party (Komma Proodevtikon)	1.95	1
KO.DE.SO. (Party of Democratic Socialism)	4.17	1
Others	4.23	

17 June 1984 European Election

	% of Votes	Seats
PASOK	41.59	10
New Democracy (Nea Democratia)	38.04	9
Communist Party of Greece (Kommounistiko Komma tes Elladas)	11.64	3
Communist Party of the Interior (Kommounistiko Komma Esoterikou)	3.42	1
E.P.EN. (National Patriotic Union)	2.28	1
Others	3.03	

Appendix

2 June 1985 Election

	% of Votes	Seats
PASOK	45.82	161
New Democracy (Nea Democratia)	40.84	126
Communist Party (Kommounistiko Komma tes Elladas)	9.89	12
Communist Party of the Interior (Kommounistiko Komma Esoterikou)	1.84	1
Others	1.61	

Bibliography

Books and Articles

Many of the Greek sources listed here and below will be found to be somewhat incomplete – as they often lack such information as publishers, dates or place of publication. This is regrettable but is a problem over which the author of this manuscript has no control.

Aggelopoulos, Aggelos. *Economikes Apopses: 1946-1958* (Economic Perspectives: 1946-1958). Athens: Difros, 1958.
Alford, Robert. *Party and Society: The Anglo-American Democracies.* Chicago: Rand McNally, 1963.
Almond, A. Gabriel and Powell, G. Bingham. *Comparative Politics: A Developmental Approach.* Boston: Little Brown and Co., 1966.
Amnesty International. *Torture in Greece: The First Torturer's Trials 1975.* London: Amnesty International Publications, 1977.
Ambler, S. John. (ed.) *The French Socialist Experiment.* Philadelphia: Institute for the Study of Human Issues, 1985.
Androulakis, Demetres. "Ta Kommata kai oi Stratigikes ton Taxeon" (Parties and the Strategies of Classes). *Kommunistike Epitheorise* (Communist Review). March 1978.
Arab Ba'th Socialist Party. *The Ba'th Party: Some Basic Considerations.* Madrid, 1977.
Asynechia Editions. *Elleniko Panepistemio Foitetiko Kinema 1983* (Greek University Students Movement 1983). Athens: Asynechia, 1983.
Athanasiades, I.T. (ed.) *24 Jouliou 1974: E Epistrophe ste Democratia kai ta Problemata tes* (24 July 1974: The Return to Democracy and its Problems). Athens: Estia, 1975.
Athenaios, Andreas. *Peripheriakes Anisotetes kai Eklogike Synperiphora* (Regional Inequalities and Electoral Behavior). Thessalonike: Kyriakides Publications, 1983.
Avdelidis, P. *E Agrotike Economia* (The Agricultural Economy). Athens: Gutenburg, 1976.
Babanasi, S. and Soula, C. *E Ellada sten Peripheria ton Anaptygmenon Choron* (Greece on the Periphery of Developed Countries). Athens: Themelio, 1978.
Babas, Stelios. "Te Enai to PASOK" (What PASOK is) Mimeo. 1974.
Bakoyiannes, Pavlos. *E Anatomia tes Ellenikes Politikes* (Anatomy of Greek Politics). Athens: Papazisis, 1977.

Bibliography

———— "Kommata sten Antiprosopeftike Democratia" (Parties in Representative Democracy) *Syntagma*. Vol. 2. 1976.
Barnet, Richard. *Intervention and Revolution*. New York: The American Library, 1972.
Becket, James. *Barbarism in Greece*. New York: Walker and Company, 1970.
Benas, D. *E Esroe tou Xenou Kefalaiou sten Ellada* (The Influx of Foreign Capital into Greece). Athens: Papazisis, 1976.
Bendix, R. *Nation Building and Citizenship*. New York: Anchor Books, 1969.
Berger, S. "Politics and Antipolitics in Western Europe in the Seventies." *Daedalus*, Winter, 1979.
Bertero, Bruno. "Structure and Problems of, and Prospects for, the Greek Economy in the 1980s." *European Cooperative* No. 23, 1980.
Bitsios, Demetres. *Pera apo ta Synora 1974 – 1977* (Beyond the Borders 1974 – 1977). Athens: Estia, 1983.
Borella, Francois. *Ta Politika Kommata tes Evropes ton Deka* (The Political Parties of the Ten European Nations). Athens: Malliares-Paedea Editions, 1983. Also available in French *Le Partis Politiques dans l'Europe des Neufs*. Paris: Editions du Seuil, 1979.
Botsaris, Demos. *Apokalypto tous Korypheous tou PASOK* (I Reveal the Top People in PASOK) Athens: Isocrates, 1983.
———— *E Dynastea Papandreou* (The Papandreou Dynasty). Athens: Isodrates, 1983.
Bourque, Gilles. "Class, Nation and the Parti Quebecois." *Studies in Political Economy* No. 2. Autumn, 1982.
Brodie, Janine M. and Jenson Jane. *Crisis Challenge and Change: Party and Class in Canada*. Toronto: Methuen Publications, 1980.
Brown, James. "Challenges and Uncertainty: NATO's Southern Flank (Political Problems in Greece and Turkey)." *Air University Review*, May – June 1980.
———— "Greek Civil – Military Relations: A Different Pattern." *Armed Forces and Society*, Spring, 1980.
Campbell, J.K. *Honour, Family and Patronage*. Oxford: Oxford University Press, 1974.
Campbell, J. and Shepard, S.P. *Modern Greece*. London: E. Benn Ltd. 1968.
Candilis, W.O. *The Economy of Greece*. New York: Praeger Press, 1968.
Carey, J.P.C. and Carey A.G. *The Web of Modern Greek Politics*. New York: Columbia University Press, 1968.
Charalabides, Michalis. *Yia ten Aftodiamorphose, Epanathemeliose tes Ellenikes Aristeras* (On the Self-Shaping, Redefinition of the Greek Left). Athens: Stochastes, 1983.
Charalambis, Demitres. *Stratos kai Politike Exousia* (Military and

Bibliography

Political Power) Athens: Exantas, 1985.

────── "Apo to Metemfiliako Kratos sto Kratos tes Metapolitefses" (From the Post-Civil war State to the Metapolitefse State) *Synchrona Themata*, July-September, 1986.

Charcavellas, T. *O Andreas Papandreou Akrovates* (Andreas Papandreou the Acrobat) Athens: Apan, 1984.

Chondrokoukis, Demetres. *E Atheate Plevra tou PASOK* (The Invisible Side of PASOK) Athens: Isokrates, 1984.

Choraphas, Vaggelis. "To Efoplistiko Kephaleo sten Ellada" (Shipping Capital in Greece). *Antithesis*15, October, 1983.

Clogg, Richard. "Greece, the End of Concensus Politics." *World Today*. London, May 1978.

────── (ed.) *The Struggle for Greek Independence*. London: Macmillan, 1973.

Clogg, Richard and Yannopoulos, George (editors). *Greece Under Military Rule*. London: Chatto and Windus, 1972. Also available in Greek *E Ellada Kato apo to Stratiotiko Zygo*. Athens: Papazisis, 1976.

Coufoudakis, V. "The European Economic Community and the 'Freezing' of the Greek Association 1967-1974." *Journal of Common Market Studies*, December 1977.

Couloumbis, A.T., Petropoulos A.J. and Psomiades, J.H. *Foreign Interference in Greek Politics* New York: Pella, 1976.

Couloumbis, T. and Iatrides, J. *Greek – American Relations: A Critical Review*. New York: Pella, 1980.

Couloumbis, T.A. *Greek Political Reaction to American and NATO Influences*. New Haven: Yale University Press, 1966.

Coutsoumaris, G. *Finance and the Development of Industry: The Greek Case*. Athens: IOBE, 1976.

────── *E Morphologia tes Ellenikes Viomichanias* (The Morphology of Greek Industry). Athens, Greek Centre of Economic Research, 1963.

Dahrendorf, Ralf. *Class and Class Conflict in Industrial Society*. Stanford: Stanford University Press, 1977.

Daphne, G. *Ta Ellenika Politika Kommata: 1821-1921* (The Greek Political Parties: 1821-1921). Athens, 1961.

Danopoulos, C. "From Balconies to the Tanks: Post-Junta Civil- Military Relations in Greece. *Journal of Political and Military Sociology.*" Vol. 13, Spring 1985.

Danopoulos, C. and Patel, K. "Military Professionals as Political Governors: A Case Study of Contemporary Greece." *Western European Politics* 3, May 1980.

Deane, Philip. *I Should Have Died*. New York: Atheneum, 1977.

Demetre, T. "E Diaspase tou '76 sto PASOK" (The Split of '76 in PASOK) *Fylladio* 6, August-October 1983.

Bibliography

Demou, Nicos. *E Chora tou Edo kai Tora* (The Country of the Here and Now). Athens: Nephele, 1983.

Demou, Tasos. *Ena Garifallo pou den t'affenoun na Anthise* (A Carnation that is not Allowed to Bloom). Athens: Glaros, 1982.

Dertilis G. *Social Change and Military Intervention in Politics* Sheffield: University of Sheffield, Unpublished Ph.D. Thesis. 1976.

Di Palma, Giuseppe. "Founding Coalitions in Southern Europe: Legitimacy and Hegemony." *Government and Opposition* 15, Spring 1978.

Diamandouros, N. "The 1974 Transition from Authoritarian to Democratic Rule in Greece: Background and Interpretation from a Southern European Perspective." Bologna: Johns Hopkins University (Occasional Papers No. 37) 1981.

Diamandouros, N.P., Kitromelides, P.M., and Mavrogordatos, G.T. (editors). *Oi Ekloges tou 1981* (The 1981 Elections). Athens: Estia, 1984.

Diakoyiannes, Kyriakos. *Politikes Skepses: Antistasiaka Keimena* (Political Thoughts: Resistance Texts). Athens: Ladias, 1975.

Dimitriou, P. *E Diaspase tou K.K.E.* (The Split of the K.K.E. – Commununist Party of Greece) 2 Volumes. Athens: Politica Provlemata, 1975.

Dovas, Yiannes. *Economia tes Elladas* (The Economy of Greece) Athens: Synchrone Epoche, 1980.

Downs, Anthony. *An Economic Theory of Democracy.* New York: Harper, 1975.

Dragasakis, Yiannes. *Yia Mia Pragmatike Allage sto Demosio Tomea tes Economias* (For a Real Change in the Public Sector). Athens: Synchrone Epoche, 1983.

Drakatos, G.K. *Ellenikes Economikes Statistikes* (Greek Economic Statistics). Athens: Papazises, 1982.

Drettakis, Manolis. *Vouleftike Ekloge 1974, 1977, 1981: Sygri-tike Melete* (Parliamentary Elections 1974, 1977, 1981: A Comparative Study). Athens, 1982.

―――― "Mia Prote Analyse ton Eklogon se Peripheriake Vase" (A First Analysis of the Elections on a Regional Basis). *Eleftherotypia*, 21 December 1977.

Duverger, Maurice. *Political Parties.* New York: John Wiley and Sons Inc., 1963.

Economic Observer. "Greece: A Case of Neocolonialism." *Monthly Review* Vol. 24 No. 7, December 1972.

Economiko Tmema tes K.E. tou K.K.E. (Economic Division of the Central Committee of the Communist Party of Greece). *E Aletheia yia ten EOK* (The Truth About the EEC). Athens: Synchrone Epoche, 1979.

Eleftheriou, Roula. "PASOK: Apo ten Auto-Organose sten Graphio-

Bibliography

cratea" (PASOK: From Self-Organization to Bureaucracy). *Anti* 241, 16 September 1983.

───── "Ennea Chronia PASOK" (Nine Years of PASOK). *Anti* 242, September 30, 1983.

Eleftheriou, S. "Ellenikos Kapitalismos: Provlemata kai Dellemata" (Greek Capitalismos: Problems and Dilemmas). *Tetradea* No. 2-3, Fall, 1981.

Elephantis, Aggelos. *E Epaggelia tes Adynates Epanastases* (The Promise of the Impossible Revolution). Athens: Themelio, 1979.

Elephantes A. and Kavouriaris M. "PASOK: Populismos e Socialismos" (PASOK: Populism or Socialism). *Polits*, October, 1977.

Eudes, Dominique. *The Kapetanios*. New York: Monthly Review Press, 1972. Also available in Greek *Oi Kapetanioi*. Athens: Exantas, 1975.

Farakos, G. *Provlema tou Ellinikou Kratikomonopoliakou Kapitalismou* (Problems of Greek State Monopoly Capitalism). Athens: Synchrone Epoche, 1976.

Fatouros, A. D. *Pera apo ten Entaxe* (Beyond Annexation). Thessalonike: Parateretes, 1980.

Featherstone, Kevin. "The Greek Socialists in Power". *Western European Politics*. July 1983.

───── "Elections and Parties in Greece (1974 – 81). *Government and Opposition* 17, Spring 1982.

───── "Socialism in one Country". *Economist*. 3 July 1982.

Filias, Vasiles. *Koinonia kai Exousia Sten Ellada* (Society and Power in Greece). Athens: Synchrona Keimena, 1974.

Finley, G. *E Istoria tes Ellinikes Epanastases* (History of the Greek Revolution) Athens, 1954.

Fisher, L. Stephen. "The 'Decline of Parties' Thesis and the Role of Minor Parties", in Merkl, F. P. (ed.) *Western European Systems: Trends and Prospects*. New York: The Free Press, 1980.

Gavilonic, Mladen. "Greek Socialists on Mount Olympus." *Review of International Affairs*. 32, No. 5, 1981.

───── "Changes at the Political Top in Greece". *Review of International Affairs* 31, June 1980.

Gellner, I. and Waterburg, J. (editors). *Patrons and Clients in Mediterranean Societies*. London: Duckworth, 1977.

Gerth, H. and Mills, C.W. *From Max Weber*, New York: Oxford University Press, 1958.

Genevoix, M. *The Greece of Karamanlis*. London: Doric Press, 1973. Also in Greek *E Ellada tou Karamanli*. Athens: Sideris, 1972.

Georgoulas, Babes. *Oi Nothes Ekloges tou 1961* (The Biased Election of 1961). Athens, 1975.

Gianaris, Nicholas. *The Economies of the Balkan Countries: Albania, Bulgaria, Greece, Romania, Turkey, and Yugoslavia*. New York:

Bibliography

Praeger, 1982.
Giannaros, G. "To Xeno Kefalaio Sten Ellenike Economia" (Foreign Capital in the Greek Economy) in E. Iliou et al., *Polyethnika Monopolia* (Multinational Monopolies). Athens: Exantas, 1973.
Gilseman, M. "Against Patron – Client Relations" in Gellner, E. and Waterbury, J. (editors) *Patrons and Clients in Mediterranean Societies*. London: Duckworth, 1977.
Gines, Yiannes. *Sten Kolase tou Parakratous* (In the Hell of the Parastate). Athens: Th. Niarchos, 1982.
Gitlin, Todd. "Counter-Insurgency: Myth and Reality in Greece" in Horowitz, David (ed.). *Containment and Revolution*. Boston: Beacon Press, 1969.
Goulielmos, Panayiotes. *O Synchronos Cosmos kai e Prooptike Tou* (The Modern World and its Perspective). Athens: 1971.
Gramsci, Antonio. *Selections from Prison Notebooks*. New York: International Publishers, 1978.
Gregoriadis, Solon. *E Istoria tes Diktatorias* (The History of the Dictatorship). 3 Volumes. Athens: Kapopoulos, 1975.
——— *Economike Istoria tes Synchrones Elladas* (Economic History of Modern Greece). Athens: 1974.
Hartwell, M.R. "The Growth of Services in the Modern Economy" in Cipolla, C. (ed.) *The Industrial Revolution*. London: Fontana, 1973.
Holden, D. *Greece Without Columns*. London: Faber and Faber, 1972.
Ioannides, Lakis. *Yiati Dialexame ton K. Karamanlis* (Why We Chose K. Karamanlis). Thessalonike: 1976.
Ionescu, Ghita and Gellner, Ernest. (eds.) *Populism: Its Meaning and National Characteristics*. London: The Mac- millan Co., 1969.
Jecchinis, Chris. *Trade Unionism in Greece*. Chicago: Labour Education Division, Roosevelt University, 1967.
Kaftantzoglou, Ioanna. *Politicos Logos kai Ideologia* (Political Logos and Ideology). Athens: Exantas, 1979.
Kamm, Henry. "Papandreou – The Politics of Anti-Americanism" *The New York Times Magazine*. April 7, 1985.
Kamouzes, Konstantine. *Anatomia tes Ellenikes Economias apo to 1953 Mechri Simera* (Anatomy of the Greek Economy from 1953 until Today). Thessalonike: University Studio, 1981.
Kallias, K. *E Ideologia tes Nea Democratia* (The Ideology of New Democracy). Athens: 1976.
Kanellopoulos, Athanasios. *E Oikonomia Anames Sto Ktes kai Avrio* (The Economy Between Yesterday and Tomorrow). Athens: Kaktos, 1980.
Kanellopoulos, Fotis. *Andreas G. Papandreou Mila ston F. Kannellopoulos yia ten Politike kai yia to Socialismo* (Andreas G. Papan-

dreou talks to F. Kanellopoulos about Politics and Socialism). Athens: 1981.
Kapetanyiannes, Vasiles. "Laikismos: Synoptikes Semioses yia mia Kritike Epanexetase" ("Populism: Comprehensive Notes for a Critical Re-assesment"). *Polites*. January-March, 1986.
Karabelias, Giorgos. *Micromesea Democratia* (Lower-Middle Democracy). Athens: Kommuna, 1982.
Karageorgas, Dionysos. "The Distribution of the Tax Burden by Income Groups in Greece". *Economic Journal*, June 1974.
Karas, Elias. "Mia Prote Gnorimia me to PASOK" (A First Acquaintance with PASOK) *Anti* 13 December 1977.
―――― "Ta Eklogika Apotelesmata kai e Ellenike Aristera" (The Election Results and the Greek Left) *Anti*, 3 December 1977.
Katephores, Giorgos. "E Protovoulia Perase Sta Cheria tou Kratous" (Initiative has Passed into the Hands of the State). *Ta Nea* 8 March 1982.
Kathephores, G. *E Nomothesia ton Vavaron* (The Barbarians' Legislation). Athens: Themelio, 1975.
―――― "Greece: the Empiricist Socialism of PASOK". *The Socialist Economic Review 1983*. London: The Merlin Press, 1983.
Kathephores, G. and Zis, G. *Fascism in Greece and Why It Happened* London: League for Democracy in Greece, 1967.
Katris, John. *Eyewitness in Greece: The Colonels Come to Power* St. Louis, Mi.: New Critics Press, 1971. Also published in Greek under the title *E Gennesis tou Neofasismou sten Ellada* (The Genesis of Neofascism in Greece) Athens: Papazisis, 1974.
Katsanevas, Theodore. *To Synchrono Syndikalistiko Kinema sten Ellada* (The Contemporary Trade Union Movement in Greece). Athens: Nea Sinora, 1981. Also in English *Trade Unions in Greece: An Analysis of the Factors Determining Their Growth and Present Structure*. Ph.D. Thesis, London School of Economics, 1980.
Katsoulis, Georgios. *To Katestemeno sten Neoellenike Istoria* (The Establishment in Modern Greek History). Athens: Nea Synora, 1975.
Kentro Mesogeakon Meleton (The Centre for Mediterranean Studies) (ed.). *Metavase sto Socialismo* (Transition to Socialism). Athens, Aletri, 1981.
K.E.M.E. *Ektemeses kai Prooptikes sten Ellada*. (Estimations and Perspectives of Employment in Greece) Athens: O.A.E.D., 1983.
Kirchheimer, Otto. "The Transformation of the Western European Party Systems". Macridis, C. Roy and Brown, E. Bernard. (editors). *Comparative Politics: Notes and Readings*. Homewood, Illinois: The Dorsey Press. 1977.
Kitsikis, D. *Istoria tou Ellenotourkikou Chorou 1928 – 1973* (The History of the Greek – Turkish Area 1928 – 1973). Athens: Estia,

Bibliography

1981.

Kitsikis, Dimitri. "Greek Communists and the Karamanlis Government". *Problems of Communism* 26. January – February, 1977.

Kohler, B. *Political Forces in Spain, Greece and Portugal.* London: Butterworth, 1983.

Kolmer, Kostas. "E Ellenike Economia se Krisimo Stavrodromi" (The Greek Economy at a Critical Crossroad) in Kentro Polikes Erevnes kai Epimorphoseos (The Centre for Political Research and Education). *To Mellon ton Men Kolliktivistikon Thesmon sten Evrope kai stis EPA* (The Future of the Non-collectivist Institutions in Europe and in the USA) Athens: Kentro Politikes Erevnes kai Epimorphoseos, 1981.

Kontogiorges, D. G. (ed.) *Koinonikes kai Politikes Dynames sten Ellada* (Social and Political Forces in Greece). Athens: Exantas, 1977.

Kordatos, Yiannes. *E Istoria tou Ergatikou Kinematos sten Ellada* (History of the Greek Workers' Movement) Athens: Boudoumanis, 1972.

——— *Selides apo ten Istoria tou Agrotikou Kinematos sten Ellada* (Passages from the History of the Peasant Movement in Greece). Athens: Boukoumanis, 1964.

——— *Esagoge es ten Istorian tes Ellenikes Kepholeokratias* (Introduction to the History of Greek Capitalism). Athens: Epikeroteta, 1977.

Korises, Chariton. *To Aftarchiko Kathestos: 1967 – 1974* (The Authoritarian Regime: 1967 – 1974). Athens: Gutenberg, 1975.

——— *E Politike Zoe sten Ellada: 1821 – 1910* (Political Life in Greece: 1821 – 1910). Athens: 1974. Also in German *Die Politischen Parteren Griechelands. Ein Neuer Staat auf dem Weg zur Demokratie 1821 – 1910.* Hersbruck/Numberg: Verlag Karl Pheiffer, 1966.

Kostopoulos, Soteres. *Yia Ena Patriotiko Socialistiko Kinema* (For a Patriotic Socialist Movement). Athens: Karanases, 1974.

——— "Apo to PAK sto PASOK" (From PAK to PASOK) *Exormise* 5 September 1975.

Kotzias, Nikos. *O "Tritos Dromos" tou PASOK* (The Third Road of PASOK) Athens: Synchrone Epoche, 1985.

Kouloglou, Stelios. *Sta Ichne tou Tritou Dromou* (At the Footsteps of the Third Road) Athens: Odysseas, 1986.

Koutsoukis, Kleomenis. *Political Leadership in Modern Greece: Cabinet Elite Circulation and Systemic Change (1946-1976)* Athens: Athena Publications S.A., 1982.

Kousoulas, G.D. *Revolution and Defeat: The Story of the Greek Communist Party.* London: Oxford University Press, 1965.

Kremmidas, G. *Oi Anthropoi tes Chountas Meta ten Diktatoria* (The Men of the Junta after the Dictatorship). Athens: Exantas, 1984.

Bibliography

Kremmidas, V. *Introduction to the History of Modern Greek Society.* Athens: Exantas, 1976.

Laclau, Ernesto. *Politics and Ideology in Marxist Theory.* London: Verso Editions, 1979.

Labrinou, G. *E Monarchia sten Ellada* (The Monarchy in Greece) Athens: Byron, 1973.

Ladis, Fondas. "Democratic or Socialist Revolution in Greece?" *Monthly Review* 24, No. 7. December 1972.

Lambiri-Dimaki, Jane. *Social Stratification in Greece 1962-1982* Athens: Sakkoulas, 1983.

Legg, R. Keith. "Political Change in a Clientelistic Polity: The Failure of Democracy in Greece". *Journal of Political and Military Sociology* 1. Fall 1973.

—— *Politics in Modern Greece.* Stanford: Stanford University Press, 1968.

Lerner, D. and Robinson, D.R. "Swords and Ploughshares: The Turkish Army as a Modernizing Force" in Bienen, H. (ed.) *The Military and Modernization.* Chicago: Aldine, 1971.

Les Temps Modernes. *E Ellada se Exelixe* (Greece in Development) Athens: Exandas, 1986.

Leys, Colin. "The 'Overdeveloped' Post-Colonial State: A Re-evaluation". *Review of African Political Economy.* January – April 1976.

Li Causi, Luciano. "Anthropology and Ideology: The Case of Patronage in Mediterranean Societies". *Radical Science Journal* 1975, No. I.

Linardatos, Spyros. "Pou Peyainei to PASOK?" (Where is PASOK Going?). *Ellenike Aristera* 11 -12. June – July 1977.

—— *E Exoterike Politike tes 4es Augoustou* (The Foreign Policy of the 4th of August Regime). Athens: Dialogos, 1975.

Linardou-Rilmon P. "Istorike e Ptose Paragòges kai Paragogetotetas kai Ependynseon ste Viomechania Kata to 1982"(Historically Significant the Drop in Productivity, Production and Investments in Industry During 1982). *Oikonomikos Tachidromos.* 16 June 1982.

—— "Entone Afxise tes Anergias Sto A'Trimeno tou 1983" (Intense Increase in Unemployment in the First Three Months of 1983). *Oikonomikos Tachidromos* June 16, 1983.

Lins, Juan. "Europe's Southern Frontier: Evolving Trends Toward What?. *Daedalus* 108, No. 1. Winter 1979.

Lipset, M. S. and Rokkan, S. *Party Systems and Voter Alignments: Cross-national Perspectives.* New York: Free Press, 1957.

Loukakos, P. "Poies Apolies tes Neas Democratias tha Epireasoun tes Mellontikestes Politikes?" (Which losses of the New Democracy Party Will Influence its Future Policies?) *To Vema.* 29 November 1977.

Loulis, John C. *The Greek Communist Party, 1940 – 1944.* London:

Bibliography

Croom Helm, 1982.

Lyrintzis, Christos. "The Rise of PASOK: The Greek Election of 1981" *West European Politics* Vol. 5. July 1982.

────── *Between Socialism and Populism: The Rise of the Panhellenic Socialist Movement* Unpublished Ph.D. Thesis, London School of Economics, 1983.

Macpherson, C.B. *Democratic Theory: Essays in Retrieval.* London: Oxford University Press, 1973.

Macridis, Roy. (ed.) *Political Parties* New York: Harper and Row, 1967.

────── "Greek Foreign Policy" Reality, Illusions, Options" in Tsoukalis, Loukas. *Greece and the European Community* Farnborough" Saxon House, 1979.

────── "To 'Symplegma tou Sotera' exothei to PASOK sten Apolytarchia" (The 'Saviour Complex' Forces PASOK to Totalitarianism). *Kathemerine.* 24-25 October 1982.

────── *E Ellenike Politike sto Stavrodromi* (Greek Politics at the Crossroads). Athens: Ellenike Evroekdotike, 1985.

Magri, Lucio. *Towards a Revolutionary Marxist Party.* Athens: Sinchrona Kemena, 1974. First published as an article in *New Left Review* No. 60, March – April 1970.

Mahaira-Odoni, Eleni. "Greece out of the Common Market?" *Telos.* No. 43, Spring, 1980.

Maltezou, Sophia. *Pios Pistevei ton Andrea?* (Who Believes in Andreas) Athens: Ermias, 1977.

Malios, M. *E Synchone Phase Anaptyxes tou Kapitalismou sten Ellada* (The Present Phase of Capitalist Development in Greece). Athens: Syncrone Epoche, 1975.

Manesis, A. and Papadimitriou, G. *The Constitution of 1975.* Thessalonike/Athens: 1976.

Maronites, Demetre. "Ellenike Kleronomia kai Europaike Kedemonia" (Hellenic Heritage and European Hegemony). *E Avge.* June 1984.

Massavetas, Giorgos. *Ste Mache yia ten Allage* (In the Battle for Change). Athens: Vasdekis, 1981.

Mavrogordatos, G. "E Laike Vase ton Kommaton kai e Taxike Syngrouse sten Ellada tou Mesopolemou" (The Popular Base of Parties and Class Conflict in Greece in the Interwar Period) *Epitheorise Koinonikon Erevnon* (The Greek Review of Social Research). No. 28, 1976.

Mavros, Giorgios. *Ethnikoi Kindinoi* (National Dangers). Athens: Atlantis, 1978.

Maximou, Serafeim. *E Avge tou Ellenikou Kapitalismou* (The Dawn of Greek Capitalism). Athens: Stochastes, 1973.

Mercouri, Melina. *I was born Greek.* London: Hodder and Stoughton,

Bibliography

1971.

Merkl, H. Peter. "Introduction: The Study of Party Systems" in Merkl, H. P. (ed.) *Western European Systems: Trends and Prospects.* New York: The Free Press, 1980.

────── "The Sociology of European Parties: Members, Voters and Social Groups" in Merkl, H. P. *Western European Systems: Trends and Prospects.* New York: The Free Press, 1980.

Metaxas, B. N. *The Economics of Tramp Shipping.* London: Athlone Press, 1971.

Meynaud, Jean. *Politikes Dynames sten Ellada: E Vasilike Ektrope apo ton Koinonvouleftesmo* (Political Power in Greece: The Royal Deviating from Parliamentarism). Athens: Byron, 1974.

────── *Oi Politikes Dymanes sten Ellada* (Political Forces in Greece). Athens: Byron, 1974. Also available in English and French.

Michail, S. *To A! Synedrio tou PASOK kai O Marxismos* (The First Congress of PASOK and Marxism). Athens: 1984.

Michels, Robert. *Political Parties.* New York: The Free Press, 1968.

Miliband, Ralph. *Marxism and Politics.* Oxford: Oxford University Press, 1978.

────── *Parliamentary Socialism.* London: Merlin Press, 1973.

────── *The State in Capitalist Society.* London: Quartet Books, 1977.

Moskov, Kostis. *E Ethnike kai Koinonike Syneidese sten Ellada* (The National and Social Consciousness in Greece). Athens: Synchrone Epoche, 1978.

Moustakis, Giorgos. *Ellas, Ellenon Despotadon* (Greece of the Greeks and of the Bishops). Athens: Gutenberg, 1983.

Mouzelis, Nicos. "Capitalism and the Development of the Greek State" in Scase, Richard. (ed.) *The State in Western Europe.* London: Croom Helm, 1980.

────── "Capitalism and Dictatorship in Post-War Greece." *New Left Review.* No. 96, March-April 1976.

────── "Class and Clientelistic Politics: The Case of Greece". *Sociological Review.* August 1978.

────── "Greek and Bulgarian Peasants: Aspects of Their Socio-political Organization during the Inter-War Period" *Comparative Studies in Society and History.* January, 1975.

────── "Ideology and Class Politics, a Critique of Ernesto Laclau." *New Left Review.* No. 112.

────── *Modern Greece: Facets of Underdevelopment.* London: Macmillan, 1978. Also in Greek: *Neoellenike Koinonia Opse Hypanaptixes.* Athens: Exantas, 1978.

────── "Modern Greece: Development or Underdevelopment?" *Monthly Review.* December 1980.

────── "The Relevance of the Concept of Class for the Study of Modern Greece" in Diner, M. and Freidl, E. (editors) *Regional Vari-*

Bibliography

ations in Modern Greece and Cyprus: Towards a Perspective on the Ethnology of Greece. New York: Academy of Science, 1976.

—— "On the Greek Elections." *New Left Review*. No. 108, March-April, 1978.

Mouzelis, Nikos and Attalikes, Michael. "Greece" in Archer, Margaret Scotford and Giner, Salvador. (editors) *Contemporary Europe: Class, Status, and Power*. London: Weidenfeld and Nicholson, 1971.

Mylonas, George. *Escape from Amorgos*. New York: Charles Scribners and Sons, 1974.

McNall, G. Scott. "Greece and the Common Market." *Telos*. No. 43, Spring 1980.

Nairn, Tom. "The Nature of the Labour Party" in Anderson, P. and Blackburn, R. (editors) *Towards Socialism*. London: Fontana/New Left Review, 1965.

National Statistical Service of Greece. *Concise Statistical Yearbook of Greece*. Athens, 1981.

Nefeloudis, B. *Apomythopoiese me te Glossa ton Arithmon* (Demythologizing the Language of Numbers) Athens: Armos, 1973.

Negreponte-Delivane, Maria. *E Ellenike Economia* (The Greek Economy). Thessalonike: Parateretes, 1981.

Nikolinakos, Marios. *Antistase kai Antipolitifse: 1967-1974* (Resistance and Opposition: 1967-1974). Athens: Olkos, 1975. Also in German under the title *Widerstand und Opposition in Griecheland*.

—— *EOK – Ellada – Mesogeios* (EEC – Greece – Mediterranean). Athens, Nea Synora, 1978.

—— *Meletes Pano ston Elleniko Capitalismo* (Studies on Greek Capitalism). Athens: Nea Sinora, 1976.

Offe, Claus. "The Separation of Form and Content in Liberal Democratic Politics." *Studies in Political Economy*. No. 3, Spring 1980.

Panitch, Leo V. *Social Democracy and Industrial Militancy* Cambridge: Cambridge University Press, 1976.

—— "The Impass of Social Democratic Politics." *Socialist Register* 1985/1986.

Panos, S. (ed.) *Ellada kai Sosialismos* (Greece and Socialism). Athens: Kastaniotes Editions, 1982.

Papakonstantinou, Theophilaktos. *Politike Agoge* (Political Education). Athens: Kabanas Hellas, 1970.

Papandreou, Andreas. *E Eleftheria tou Anthropou* (The Freedom of Man). Athens: Karanase, 1971.

—— *Paternalistic Capitalism*. Toronto: The Copp Clark Publishing Company, 1972.

—— "The Takeover of Greece." *Monthly Review*. 24, No. 7, December 1972.

—— *Democracy at Gunpoint*. London: Penguin, 1973. Also in avail-

able in Greek: *Democratia sto Apospasma*. Athens: Karanasis, 1974.
────── *Imperialismos kai Economike Anaptyxe* (Imperialism and Economic Development). Athens: Nea Synora, 1975.
────── "Sosialismos sten Ellada, Pos kai Pote?" (Socialism in Greece, how and when?) interview in *Ta Nea*. November 1975.
────── *Apo to PAK sto PASOK* (From PAK to PASOK) Athens: Ladias, 1976.
────── "E Mesogiake Koine Agora" (Mediterranean Common Market). Interview in *Eleftherotypia*. 2 February 1976.
────── *E Ellada stous Ellenes* (Greece to the Greeks). Athens: Karanase, 1976.
────── "Greece: The Meaning of the November Uprising". *Monthly Review*. February 1977.
────── *Yia Mia Socialistike Koinonia* (Toward a Socialist Society). Athens: Aichme, 1977.
────── "A. G. Papandreou Synentefxe" (Interview with A. G. Papandreou). *Ta Nea*. 26 February, 1978.
────── "Confrontation and Coexistence". *Monthly Review*. 29, No. 11. April, 1978.
────── *Metavase sto Socialismo: Provlemata kai Strategike yia to Elleniko Kinema* (Transition to Socialism: Problems and Strategies for the Greek Movement). Athens: Aichme, 1981.
────── "O Marx, O Lenin kai e Dictatoria tou Proetariatou" (Marx, Lenin and the Dictatorship of the Proletariat). *Exormise*. November 1975. and in *Metavase sto Socialismo* (Transition to Socialism). Athens: Aichme, 1981.
Papandreou, Margaret. *Nightmare in Athens*. Englewood Cliffs, New Jersey: Prentice-Hall, 1970. Also available in Greek: *Efialtes sten Athena*. Athens: Kastaniotes, 1984.
Papandreou, Vaso. *Polyethnikes Epichereses* (Multinational Corporations). Athens: Gutenberg, 1981.
Papasarantopoulos, P. (ed.) *PASOK kai Exousia* (PASOK and Power). Athens: Paratirites, 1980.
Papaspiliopoulos, Spilios. (ed.) *Meletes Pano ste Synchrone Economia* (Studies on the Modern Greek Economy). Athens: Papazisis, 1978.
────── "The Last Twenty Years: Industrialization Shores up the Greek Economy". *European Communities*. January-February 1981.
────── "The Political Forces in the Next Elections". *Anti* November, 1977.
Paralikas, K. Dimitris. *I.D.E.A. kai A.S.P.I.D.A.: Rizes kai Plokamia* (I.D.E.A. and A.S.P.I.D.A.: Roots and Ramifications). Athens: 1978.
Paschos, Giorgos. "Ethinike Laike Enoteta kai Eklogiko Systema" (National Popular Unity and Electoral System). *Polites*. January-

Bibliography

March 1986.

Penniman, Howard. (ed.) *Greece at the Polls: The National Elections of 1974 and 1977.* Washington: American Enterprise Institute for Public Policy Research, 1981.

Pesmazoglou, Stefanos. *"Vioi Paralleloi" ston Evropaiko Noto* (Parallel Lives in the European South). Thessalonike: Paratirites, 1986.

Petras, James. "The Contradictions of Greek Socialism." *New Left Review* No. 163 May-June 1987.

────── "A Critical Appraisal of US Policy Toward Greece." Mimeo.

────── "Greece: Foreign Policy Alternatives for the 1980s." Mimeo.

────── "Notes Toward a Definition of Greek Political Economy" Mimeo.

────── "The Rise and Decline of Southern European Socialism." *New Left Review.* No. 146 July-August 1984.

────── "Oi Taxes sten Ellada" (Classes in Greece). *Monthly Review* (Greek Edition). No. 36, May, 1983.

Petropoulos, J. A. *Politics and Statecraft in the Kingdom of Greece: 1833-1843.* New Jersey: Princeton University Press, 1968.

Photopoulos, A. "The Dependence of the Greek Economy on Foreign Capital." *Economicos Tachidromos.* July 1975.

Poulantzas, Nicos. "Political Parties and the Crisis of Marxism" Interview in *Socialist Review.* No. 48 (9.6).

────── *The Crisis of Dictatorships* London: New Left Review Publications, 1976. Also available in Greek *E Crise ton Dictatorion.* Athens: Papazisis, 1977.

────── *Political Power and Social Classes.* London: Verso, 1978.

Powell, J. D. "Peasant Society and Clientelistic Politics." *American Political Science Review.* 64. 1970.

Przeworski, A. "Social Democracy as an Historical Phenomenon". *New Left Review.* No. 122. July-August 1980.

Przeworski, Adam and Sprague, John. "A History of Western European Socialism." Mimeo. (Paper presented at the Annual Meeting of the American Political Science Association, Washington, D.C., September, 1977).

Psiroukis, Nikos. *E Diamache Sto Aegeo* (The Aegean Dispute). Athens: Epikairotita, 1977.

────── *E Entaxe tes Elladas sten EOK kai o Allos Dromos* (Greece's Annexation to the EEC and the Other Road). Athens: Synchrona Themata, 1976.

────── *Istoria tes Synchrones Elladas* (History of Contemporary Greece). 4 Volumes. Athens: Epikairotita, 1976.

────── *Istorikos Choros kai Ellada* (Historical Space and Greece). Athens: Epikairotita, 1975.

Ralles, Giorgios. *Choris Prokatalepse* (Without Prejudice). Athens: Ellenike Euroekdotike, 1983.

Bibliography

———— *E Technike tes Vias* (The Technique of Violence). Athens: Ermias, 1976.
———— *Ores Efthynes* (The Times of Responsibility). Athens: Evroedotike, 1983.
Rendez-vous Me Ten Istoria (Rendez-vous With History). Athens: Cactus Editions, 1981.
Richards, J. "Populism: A Qualified Defence". *Studies in Political Economy*. No. 5. Spring 1981.
Rokkan, S. *Citizens, Elections and Parties*. New York: McKay Co., 1970.
Ross, G. and Jenson, J. "Crisis and France's 'Third Way'". *Studies in Political Economy*. No. 11. Summer 1983.
———— "Pluralism and the Decline of the Left Hegemony: The French in Power". Mimeo. June 1975.
Rozakis, Christos. *Tria Chronia Ellenikes Exoterikes Politikes: 1974 – 1977* (Three Years of Greek Foreign Policy: 1974 – 1977). Athens: Papazisis, 1978.
Rustow, A. D. "Turkey: The Modernity of Tradition" in Rye, W. L. (editors) *Political Culture and Political Development*. Princeton: Princeton University Press, 1969.
St. Martin, Katerina. *Labrakides: Istoria mias Genias* (Labrakides: the History of a Generation). Athens: Polytypo, 1984.
Samaras, James. *Kratos kai Kephalaio sten Ellada* (State and Capital in Greece). Athens: Sinchrone Epoche, 1977.
San Francisco Bay Area Kapitalistate Group. "Political Parties and Capitalist Development." *Kapitalistate*. No. 6. Fall 1977.
Sarres, Demetres. *Politika Keimena* (Political Writings). Iraklion, 1982.
Sartori, Giovanni. *Parties and the Party System: A Framework for Analysis*. Cambridge, England: Cambridge University Press, 1976.
Saul, John. "The State in Post Colonial Societies." *Socialist Register*, 1974.
Seirenides, Kostas. *E Stratiotike Parousia ton EPA sten Ellada* (The Military Presence of the USA in Greece). Athens, Synchrone Epoche, 1983.
Sellou, D. "To PASOK Meta to Synedrio" (PASOK After the Congress). *Fylladio*. No. 10. September, 1984.
Semites, Kostas. *E Domike Antipolitefse* (The Structural Opposition). Athens, Kastaniotes, 1979.
———— *Politike – Kyvernise – Dikaio* (Politics – Government – Law). Athens: Kastaniotes, 1981.
Serafetinidis, Melina. *The Breakdown of Parliamentary Democracy in Greece*. Unpublished Ph.D. Thesis, London School of Economics and Political Science, 1977.
Serafetinidis, M., Serafetinidis, G., Lambrinides, M., Demathas Z.

Bibliography

"The Development of Greek Shipping Capital and its Implications for the Political Economy of Greece." *Cambridge Journal of Economics*. 5, 1981.

Small Scale Industry Working Group. *Development Plan 1976 – 1980* Athens: Center of Planning and Economic Research, Small Scale Industry Working Group Report No. 2, 1976.

Spourdalakis, M. *Convergence of Political Parties: The Greek Case.* M.A. Thesis. University of Manitoba, 1980.

———— "Yia te Theoria ton Politikon Kommaton" (On the Theory of Political Parties). *The Greek Review of Social Research* No. 55. 1984.

———— "Towards an Understanding of Greek Politics" *Hellenic Studies*. Vol. II No. 2. Autumn 1984.

———— "The Greek Experience," *Socialist Register* 1985/86.

Stamiris, Eleni. "The Women's Movement in Greece" *New Left Review*. No. 158 July-August 1986.

Stefanakis, Giorgos. *E Glossa Tes Allages* (The Language of Change). Athens: Estia, 1983.

Svoronos, Nicos. *Analecta Neo-Ellenikes Istorias kai Istoriographias* (Analecta of Modern Greek History and Historiography). Athens: Themelio, 1982.

———— *Episkopese tes Neoellenikes Istorias* (An Overview of Modern Greek History). Athens: Themlio, 1976.

Therborn, Goran. "The Role of Capital and the Rise of Democracy" *New Left Review*. No. 103. May – June 1977.

Thomas, Clayton John. "Ideological Trends in Western Political Parties" in Merkl, H. P. (ed.) *Western European Systems: Trends and Prospects*. New York: The Free Press, 1980.

Tourkovasile-Galanou, R. *Ta Politika Kommata: Meta to 1967* (The Political Parties: After 1967). Athens: Papazises, 1980.

Tsouderos, E. J. *Agrotikoi Syneterismoi* (The Agricultural Co-operatives). Athens: Estia, 1960.

Tsoucalas, Konstantine. "Class Struggle and Dictatorship in Greece." *New Left Review*. No. 56. July-August 1969.

———— *The Greek Tragedy*. London: Penguin, 1969.

———— *Exartise kai Anaparagoge* (Dependency and Reproduction) Athens: Themelio, 1977.

———— *Koinonike Anaptexe kai Kratos* (Social Development and the State). Athens: Themelio, 1981.

———— "Koinonikes Proektases tes Demosias Ergodosias Ste Matapolemike Ellada" (Social Implications of Public Employment in Post-War Greece). *The Greek Review of Social Research* No. 50. 1983.

Tsoukalis, Loukas. *The European Community and its Mediterranean Enlargement*. London: George Allen and Unwin, 1981.

Bibliography

Tsourakis, Dionysios. *Ta Paraskenia tes Apotychias tou PASOK* (The Behind the Scenes Facts of PASOK's Failure). Athens: Ellenike Evroekdotike, 1984.

Tucker, Robert C. "The Deradicalization of Marxist Movements" in Macridis, Roy C. and Brown, Bernard E. (eds) *Comparative Politics: Notes and Readings*. Homewood, Illinois: The Dorsey Press, 1968.

Tumin, Melvin, M. *Social Stratification: The Forms and Functions of Inequality*. New Jersey: Prentice-Hall, 1967.

Tzoannos, I. G. *Ellenike Emporike Naftilia kai e EOK* (Greek Merchant Shipping and the EEC). Athens: I.O.B.E., 1977.

Vasileiades, Damianos. *PAK – PASOK: Mythos kai Pragmatikotita* (PAK – PASOK: Myth and Reality). Athens: Dialogos, 1977.

Veremis, Thanos. *Greek Security Considerations*. Athens: Papazises Publishers, 1982.

Vergopoulos, Kostas. *To Agrotiko Zetema sten Ellada* (The Agricultural Issue in Greece). Athens: Exantas, 1975.

——— "Eisagoge sten Neo-ellenike Economia" (Introduction to the Modern Greek Economy). *Monthly Review* (Greek Edition). No. 36, May 1983, No. 37, June 1983.

Vernicos, N. *E Ellada Mprosta ste Dekaetia tou '80* (Greece facing the Decade of the 80s). Athens: Exantas, 1975.

Vlanta, Demetre. *Andreas Papandreou: Epephanea kai Pragmatikotita* (Andreas Papandreou: Appearance and Reality). Athens:1978.

Voethema yia ten Istoria tou K.K.E. (Textbook on the History of the Communist Party of Greece). Athens: Koinonikes Ekdoses, 1978.

Vournas, Tasos. *Syntome Istoria tes Ellenikes Epanastases* (Brief History of the Greek Revolution). Athens: Drakopoulos Editions.

Whitaker, Reginald. "Considerations on a Political Economy of Political Parties". Mimeo. (Paper prepared for presentation to the Annual Meeting of the Canadian Political Science Association, University of Quebec at Montreal, June 2, 1980.)

Wittner, Lawrence. *American Intervention in Greece, 1943 – 1949*. New York: Columbia University Press, 1982.

Woodward, S.V. "Epikivsyna Refste kai Akathoriste e physe tou 'Politikou Laikismou'" (The Nature of 'Political Populism': Dangerously Fluid and Undefined). *Kathemerine*. September 21, 22, 1983. First Published in *New Republic*.

Yannakakis, Ilios. "The Greek Communist Party." *New Left Review*. No. 54. March-April 1969.

Yeachnes, Vaggelis. *To Telos tes Microastikes Evdaimonias* (The End of Petty-Bourgeois Eudaimonia). Athens: Exantas, 1985.

Zachareas, Aemilios. *Yia ten Enoteta tes Ellenikes Marxistikes Aristeras* (For the Unity of the Greek Marxist Left). Athens: Dialogos,

Bibliography

1974.

——— *E Aristera kai e Economike Crise* (The Left and the Economic Crisis). Thessalonike: Parateretes, 1983.

Zephyros, K. "E Ekonomike Politike tou PASOK yia to 1983" (The Economic Policy of PASOK for 1983). *Marxistiko Deltio* (Marxist Bulletin). No. 17, January – March 1983.

Ziemann, W. and Lanzendorfer. "The State in Peripheral Societies". *Socialist Register.* 1977.

Zapheropoulos, Yiannes. *E Nea Kyvernitike Dome* (The New Governmental Structure). Athens: Epalxe Publications, 1982.

Zolotas, X. *Monetary Equilibrium and Economic Development.* Athens:1964.

Zorbalas, Stavros. *O Neofasismos sten Ellada: 1967 – 1974* (Neofascism in Greece: 1967 – 1974). Athens: Synchrone Epoche, 1978.

Party Documents

PASOK

Apofases tes Prosorines Kentrikes Epitropes tou PASOK Yia ten Anaptexe tes Organoses tou Kinematos (Resolution of the Provisional Central Committee of PASOK – For the Advancement of the Organization of the Movement). No precise date available. Approximately December 1974.

Apofases tes Synelefses tou PASOK Italias (Resolution of the Conference of PASOK of Italy). Mimeo. No precise date available. Winter 1975.

Apofases tou PASOK Italias (Resolutions of the Italian Sections of PASOK). Mimeo. 29-30 November 1975, 11 December 1976.

A' Synedriose tes Synelefses tou PASOK: 29-8-1979 (The First Meeting of the Conference of PASOK on 29 August 1976). Athens: Publications Committee – PASOK, 1976.

Brosta stes Ekloges kai te Nike Synedriase tes 8es Synodou tes Kentrikes Epitropes tou PASOK (Facing the Elections and the Victory – The Meeting of the 8th Session of the Central Committee of PASOK). Athens: Publications Bureau, KE.ME.DIA./PASOK, 1981.

Diakeryxe Kyvernitikes Politikes: Symbolaio me to Lao (Declaration of Governmental Policies: Contract With the People). Athens: Publications Bureau, KE.ME.DIA./PASOK, 1981.

Dokoumenta tes A' Panelleneou Synedriou tes PASP (Documents of the A' Panhellenic Congress of PASP). Athens, 1976.

Bibliography

E Apele sto Aegeo (The Threat in the Aegean). Athens: Publications Bureau, PASOK, 1977.

E Economike Politike Sten Ellada meta ton B' Pankosmio Polemo (Economic Policy in Greece after the Second World War). Athens: Agonistes Editions, 1975.

Eirene – Apopyrenikopoiese (Peace – Disarmament). Papandreou's Speech to the European Congress for a Nuclear Free Europe. Athens: Publications Bureau, KE.ME.DIA.

Eisegeses kai Omilies tou Proedrou (Proposals and Speeches of the President). Athens: November 1975.

Ellada kai Koine Agora: O antilogos (Greece and the Common Market: The Counter-Argument). Athens: November 1976.

Ethnike Laike Enoteta Synedriase tes 2es Synodou tes Kentrikes Epitropes tou PASOK (National Popular Unity – The Meeting of the Second Session of the Central Committee of PASOK). Athens: Publications Bureau, KE.ME.DIA., 1978.

E These tes Gynaikas sten Koinonia (The Place of Woman in Society) Athens: Publications Bureau, KE.ME.DIA./PASOK, 1976.

Ethnike Anexartesia, Laike Kyriarchia, Koinonike Apoleftherose kai Democratike Diadikasia (National Independence, Popular Sovereignty, Social Liberation and Democratic Procedure). Athens, June 1976.

Greek Government Programme. Presented to Parliament by the Prime Minister Andreas G. Papandreou. Athens: General Secretariat for Press and Information, 1981.

Hypoepitrope Syndikalistikou yia tous Ergazomenous (Trade Union Sub-Committee on Workers). Athens: Trade Union Subcommittee, No precise date available, approximately 1975.

Kanonismoi Leitourgias kai Monimes Egyklioi (Regulations for Operation and Permanent Memos). Athens: Publications Bureau, KE.ME.D.I.A./PASOK, 1984.

Keimena. Seires A', Volume 2. Athens: Publications Bureau, June 1977.

Koinonikes Taxes kai PASOK (Social Classes and PASOK). Athens: May 1975.

Koinonikopoiese. Anakoinose tou Prothepourgou A. Papandreou. (Socialization – Announcement of the Prime Minister A. Papandreou). Athens: 1983.

Laos – PASOK sten Exousia. Omilies tou Proedrou tou PASOK ste Kalabaka, Alexandroupole, Sparta. (People – PASOK in Power – Speeches of the President of PASOK in Kalabaka, Alexandroupole, Sparta). Athens: Publications Bureau, KE.ME.D.I.A., 1981.

Laos – PASOK sto Dromo tes Allages. Synedriase tes 9es synodou tes Kentrikes Epitropes tou PASOK (People – PASOK on the Road to

Bibliography

Change – The Meeting of the 9th Session of the Central Committee of PASOK). Athens: Publications Bureau, KE.ME.D.I.A./PASOK, 1982.

Letter from the "Expelled and Administratively Removed Members of PASOK to the Executive Bureau". Mimeo. October 1976.

30 Menes: Themelionoume me Erga ten Allage (30 Months: We Have been Founding Change with Actions). Athens: PASOK Editions, 1984.

Memo of the Trade Union Bureau. Report and Proposals. Mimeo. 1974, No exact date available.

Meniaio Enemerotiko Deltio (Monthly Information Bulletin). 1976-1984.

Notaras, G. *Proposal on the Constitution of the Central Committee.* Mimeo. No precise date available, approx. December 1974.

Oi Theses Mas: PASP (Our Positions: Panhellenic Militant Students' Front). Athens: no publisher or date available.

Oi Theses tes K.E. tou PASOK yia to Synedrio (The Thesis of the C.C. of PASOK for the Congress). Athens: Publications Bureau, KE.ME.D.I.A./PASOK, 1984.

Papandreou, A. *Apophasistikes Stigmes* (Decisive Moments). Athens: November 1975.

―――― *Common Market.* Athens: PASOK Editorial Committee, 1978.

―――― *Edafike Akeraioteta kai Exoterike Politike* (Territorial Integrity and Foreign Policy). Athens: Publications committee, PASOK, 1976.

―――― *Eisegeses yia Zymose* (Proposals for Zymosis). Selection of Speeches. Athens: Publications Committee, 1978.

―――― *E Istorike Anametrise: Synetefxes yia tes Ekloges* (The Historical Confrontation: Interviews on the Elections) Athens: Publications Bureau, KE.ME.DIA., 1981.

―――― *Omilies stous Epistemones kai Kallitechnes* (Speeches to the Scientists and Artists). Athens: Publications Bureau, KE.ME.DIA., 1981.

―――― *Poria Pros ten Allage: Omilies tou Proedrou tou PASOK* (March Towards the Change: Speeches of the President of PASOK). Nos. 1-5. Athens: Publications Bureau KE.ME.DIA., 1978-1981.

―――― *Provlematismoi yuro apo to Neoaistiko Kinema* (Problematic Around the Youth Movement). Athens: 1979.

―――― *To Agrotiko Provlema ste Chora Mas* (The Agricultural Problem in Our Country). Selections from Papandreou's Speeches. Athens: Publications Bureau, KE.ME.DIA./PASOK 1966.

―――― *Yia Mea Mesoyeio Apeleftheromene kai Sosialistike Eisegeses stes Mesoyeiakes Syndiaskepses, 1977, 1979* (For a Liberated and

Bibliography

Socialist Mediterranean: Proposals to the Mediterranean Conferences, 1977, 1979). Athens: Publications Bureau, KE.ME.DIA./PASOK, 1981.

PASOK 1) To PASOK Kyria Dyname tes Antipolitefses. 2) To 1976 na Chtisoume to PASOK pio Dynato. 3) Oloi ste Douleia yia ena pio Maziko kai Dynato PASOK (1. PASOK the Main Force of the Opposition. 2. To Build a Stronger PASOK in 1976. 3. All to Work for a Mass and Strong PASOK). Athens: January 1976.

PASOK Positions. Opening Speech of Andreas Papandreou in the First Post-Election Parliamentary Session – December 15, 1977. Athens: Edition Committee, 1978.

Philippatos, Lefteris. "Gramma ston Proedro" (Letter to the President). Mimeo. 5 March, 1975.

Polydiastate Exoterike Politike. Synentefxes tou Protheporgou (Multidimensional Foreign Policy – Prime Minister's Interviews). Athens: General Secretariat for Press and Information, Information Division, 1982.

Pos Organonetai to PASOK (How PASOK Organizes Itself). No place, publisher or date available.

Provlemata Taktikes tou PASOK (PASOK's Problems on Tactics) Athens: No exact date available, approximately 1975.

Sto Dromo tes Allages (On the Road to Change). Athens: 1982.

Syntagma Yia Mia Ellada Democratike (Constitution for a Democratic Greece). Athens: Kedros, 1975.

Ta Kathikonta ton Melon tou PASOK: E Kataktise ten Socialistike Synedise (The Duties of PASOK's Members: Conquering the Socialist Consciousness). Athens: KE.ME.DIA., 1977.

Ti Thelei to PASOK (What does PASOK Want). Athens: KE.ME.DIA. Publication, No exact date available, likely 1974 or early 1975.

Theses tou PASOK yia ta A.E.I. (PASOK's Thesis on Secondary Education). Athens: Publications Committee PASOK, 1978.

To PASOK sten Kyvernes O Laos sten Exousia Synedriases 6es kai 7es Synodou tes Kentrikes Epitopes tou PASOK (PASOK in Government, the People in Power – The Meetings of the 6th and 7th Sessions of the Central Committee of PASOK). Athens: Publications Bureau, KE.ME.DIA., 1981.

To Ergo tes Allages (The Achievements of Change). PASOK, 1985.

To Demotiko Programma tou PASOK. (The Municipal Program of PASOK). Athens: Publications Committee, PASOK, 1979.

To PASOK Sto Diethne Choro (PASOK in the International Arena). Athens: Publications Committee, 1978.

Yia Mia Mazike kai Protopora Organose. Synedriase tes 5es Synodou tes Kentrikes Epitopes tou PASOK (For a Mass and Vanguard Organization – The Meeting of the 5th Session of the Central Commit-

tee of PASOK). Athens: Publications Bureau KE.ME.DIA./ PASOK, 1979.
Yia ten EOK. Synentefxes kai Deloses tou Proedrou tou PASOK (On the EEC – Interviews and Declarations of the President of PASOK). Athens: Publications Bureau, KE.ME.DIA., 1981.

Nea Democratia (New Democracy)

Karamanlis, Konstantine. *Omilies kai Deloses: Ioulios 1974 - Maios 1976* (Speeches and Declarations: July 1974 – May 1976). Athens: Publication of the Party of New Democracy, No date available.
Nea Democratia: Eklogikon Programma (New Democracy: Electoral Programme).
Nees Theses: Analysis Ideologikon Archon tes Neas Democratias (New Positions: Analysis of the Ideological Principles of New Democracy). Athens: Edition of the Journal 'Nea Politike', 1976.
Politikes Theses tou Konstantine Karamanlis (The Political Positions of Konstantine Karamanlis). Athens: 1976.

K.K.E. (Communist Party of Greece)

Apophases tou 10ou Synedriou tou K.K.E. (Resolutions of the 10th Congress of the Communist Party of Greece). Athens: Publication of the Central Committee of the Communist Party of Greece, 1978.
Economiko Tmema tes K.E. tou K.K.E. (Economic Division of the Central Committee of the Communist Party of Greece). *E Aletheia yia ten EOK* (The Truth About the EEC). Athens, Synchrone Epoche, 1979.
Florakis, Charilaos. *Ekthese Drasis tes Kentrikes Epitropes sto 10o Synedrio tou K.K.E.* (Report of Action by the Central Committee at the 10th Congress of the Communist Party of Greece). Athens: Publication of the Central Committee of the Communist Party of Greece, 1978.
K.K.E.: Eklogiko Programma 1977 (Communist Party of Greece: Electoral Programme of 1977).
K.K.E.: Eklogiko Programma: 1981 (Communist Party of Greece: Electoral Programme of 1981).
K.K.E.: Katastatiko (Communist Party of Greece: Constitution). Athens: Publication of the Central Committee of the Communist Party of Greece, 1978.
Programma tou K.K.E. (The Programme of the Communist Party of Greece). Athens: Publication of the Central Committee of the

Bibliography

Communist Party of Greece, 1978.
11o Synedrio tou K.K.E.: Documenta (11th Congress of the Communist Party of Greece: Documents). Athens: Publications of the Central Committee of the Communist Party of Greece, 1982.
Theses tes Kentrikes Epitropes tou K.K.E. Yia to 11o Synedrio (Proposals of the Central Committee of the Communist Party of Greece for their 11th Congress). Athens: 1982.

K.K.E. Esoterikou *(Communist Party of the Interior)*

K.K.E. Esoterikou: Eklogiko Programma (Communist Party of the Interior: Electoral Programme 1981).
"Prosynedriakes Theses tes Kentrikes Epitropes tou K.K.E. Esoterikou" (Pre-Congress Theses of the Central Committee of the Communist Party of the Interior). *Kommounistike Theoria kai Politika*. February – March 1976.
"Skedio Programmatos tou K.K.E. Esoterikou" (Outline for a Programme of the Communist Party of the Interior). *E Avge* 9 November 1975.

Sosialistike Poria *(Socialist March)*

Apophases tes A' Panhellenias Syndiaskepses (Resolutions of the First Panhellenic Conference). Athens: Sosialistike Poria, 1976.
Apophases tes B' Panhelladikes Syndiaskepses (Resolutions of the Second Panhellenic Conference). Athens: Sosialistike Poria, 1976.
Apophases tou Ektatou Synedriou tes Sosialistikes Porias (Resolutions of the Special Congress of Socialist March). Athens: 1978.
Sosialistike Poria: Esoterika Deltia (Socialist March: Internal Bulletins). 1976-1979.

Symmachia *(Alliance of Progressive and Democratic Forces)*

Programatikoi Stochoi tes Symmachias ton Proodeftikon kai Aristeron Dynameon (Programmatic Aims of the Alliance of Progressive and Left Forces). Athens: 1977.
Symmachia Proodeftikon kai Aristeron Dynameon: Eklogiko Programma (The Alliance of Progressive and Left Forces: The Electoral Programme). 1977.

Bibliography

Other Parties and Groups

Anexartetoi Aristeroi – Kommounistes tes Ananeoses (Independent Left – Communists of the Renaissance Left). *Yia ten Epikairotera tou Sosialismou* (On the Timeliness of Socialism). Athens: September, 1981.

Enose Demokratikou Kentrou (Union of the Democratic Centre). *Manifesto 77.*

Documents Referring to the Dictatorship Period

Agonas (Struggle). 28 July, 1973. 27 September, 1973.

Demokratike Amena (Democratic Defence). *Ideologikoi kai Polikoi Stochoi* (Ideological and Political Aims). Athens: September 1974.

Demokratike Amena (Democratic Defence). *Katastatiko* (Constitution).

KOM.EP. (Communist Review). Volumes 6/1970 and 10/1971.

Neos Kosmos (New World). Volume 10/1967. No. 3-4, March-April 1974.

O Apeleftherotikos Agonas (The Liberation Struggle). PAK 1974.

Oi Stochoi tou Agona tou PAK (The Aims of PAK's Struggle). 1973.

Papandreou, Andreas. "E Anage Exsynchronismou ton Politikon Kommaton" (The Necessity for the Modernization of Political Parties). Speech at the *Greek Heritage Symposium* October 6, 1965. Mimeo. Later Published by Centre Union in Italy in June 1967.

Stangos, Asteris. "E Anage yia ena Neou Typou Politiko Kinema" (The Necessity for a New Type of Political Movement). Proposal in the Democratic Defence Congress, August 1974. Mimeo.

Journals – Periodicals

Agonistes
Agora
Anti
Antiplerophorese
Antitheses
Blueline: Greek and Mediterranean Report
Fylladio
Polites
Spotlight (A Fortnightly Publication of the Institute for Political Studies)

Bibliography

Tetradia
Socialistike Theoria kai Proxe
Xekinema

Newspapers

Athenaike
Ta Nea
To Vema
E Avge
Eleftherotypia
Kathemerine
Mesemvrene
Rizospastes

Other Documents

The Greek Economy in Figures. Athens: Electra Press, 1984.
The Greek Economy Today. Athens: Center of Planning and Economic Research, January 1984.
Investment Incentives in Greece. Athens: General Secretariat for Press and Information, Information Division, 1982.
Investment Policies in Greece. Athens: General Secretariat of Press and Information, Information Department, 1983.
Odegos Eklogon ap' to 1961 (Electoral Guide Since 1961). Athens: Vergos Publications, 1977.

Interviews

Apostolopoulos Apostolis, Political reporter and anchorman for the National Television Network (ERT). Interview held on 20 August 1984.
Babas, Stelios. Former member of PASOK's Trade Union Bureau and former member of the Executive Secretariat of Sosialistike Poria (Socialist March). Interview held on 12 July 1983.
Notaras, Gerasimos. Prominent member of Democratic Defence, member of the Central Committee of PASOK which was elected by the Pre-Congress. Interview held on 7 July 1984.
Polites, Andreas. Member of the founding Central Committee of PASOK Interview held on 26 July 1983.

Bibliography

Stangos, Asteris. Prominent member of Democratic Defence, member of the founding Central Committee which was elected by the Pre-Congress and member of PASOK's Executive Bureau, until the summer of 1975. Interview held on 4 July 1983.

Tzafoulias, George. Former Secretary of PASOK in Florence, Italy. Interview held in 23 July 1984.

Tzouvanos, Demetres. Former member of the Central Committee of PASOK and editor of *Fylladio*. Interview held on 22 August 1983.

Vasileiades, Damianos. Former Secretary of PAK – Munich and former member of the Central Committee of PASOK. Interview held on 17 July 1984.

Yiakoumelos, Nikos. At the time of the interview, he was special advisor to the Minister of Agriculture. Interview held on 10 July 1983.